STP 1448

Technology of Floor Maintenance and Current Trends

William J. Schalitz, editor

ASTM Stock Number: STP1448

ASTM International
100 Barr Harbor Drive
PO Box C700
West Conshohocken, PA 19428-2959

Printed in the U.S.A.

Library of Congress Cataloging-in-Publication Data

Technology of floor maintenance and current trends/William J. Schalitz, editor.
p. cm.—(STP; 1448)
"ASTM stock number STP1448."
Includes bibliographical references.
ISBN 0-8031-3482-7
1. Floors—Cleaning. 2. Floor polishes—Standards. I. Schalitz, William J., 1966– II. ASTM International. III. ASTM special technical publication; 1448.

TH2528.T43 2004
667′.72—dc22

2004048854

Peer Review Policy

Each paper published in this volume was evaluated by two peer reviewers and at least one editor. The authors addressed all of the reviewers' comments to the satisfaction of both the technical editor(s) and the ASTM International Committee on Publications.

To make technical information available as quickly as possible, the peer-reviewed papers in this publication were prepared "camera-ready" as submitted by the authors.

The quality of the papers in this publication reflects not only the obvious efforts of the authors and the technical editor(s), but also the work of the peer reviewers. In keeping with long-standing publication practices, ASTM International maintains the anonymity of the peer reviewers. The ASTM International Committee on Publications acknowledges with appreciation their dedication and contribution of time and effort on behalf of ASTM International.

Printed in Haddonfield, NJ
September 2004

Foreword

This publication, *Technology of Floor Maintenance and Current Trends*, contains papers presented at the symposium of the same name held in Las Vegas, Nevada on 14 October 2002. ASTM International Committee D21 on Polishes and the International Sanitary Supply Association (ISSA) sponsored the symposium. The symposium chairman was William J. Schalitz, Spartan Chemical Company, Inc, Maumee, Ohio.

Contents

Overview

ASTM Committee D21 on Polishes is charged with the responsibility of establishing the consensus standards by which floor polish composition, performance, and safety are determined. In conjunction with this responsibility comes a requirement, or unwritten expectation, that the consuming public of these polishes be educated to a degree which allows them to not only understand the governing standards issued by Committee D21, but also make informed decisions concerning the considerable amount of "alternative" opinions they are presented with on a consistent basis.

To support the Committee's consumer outreach, D21 organized a general topics symposium, titled "Technology of Floor Maintenance and Current Trends", that was held on October 14, 2002 in Las Vegas, NV. To maximize the symposium's exposure in terms of polish users, marketers, and manufacturers, it was held in conjunction with the International Sanitary Supply Association (ISSA) annual convention. ISSA is the premier trade association for the industrial and institutional cleaning industry, with the annual convention drawing in excess of 15,000 attendees and 700 exhibitors. The results of this cooperative effort were clearly demonstrated in the fact that the symposium was well attended by a group that demographically represented exactly the target market we had hoped to reach.

The papers found in this book are not limited to those presented at the Technology of Floor Maintenance and Current Trends Symposium. Additional authors have contributed to ensure that the publication has a broad base of appeal from the formulation chemist developing polishes to the facilities manager who is looking to better manage his floor care program. From a general standpoint, the papers can be broken down into three broad categories relevant to the current state of the polish industry.

The publication gets started with a block of papers focused on the various chemistries involved in building polishes and associated coatings. The first paper provides a thorough review of floor polish chemistry and presents it in such a manner that allows even non-technical individuals the ability to better understand the dynamics associated with floor polish formulation. After a review of the chemistry involved, the subject matter turns to two very significant areas of concern: maintenance and static coefficient of friction.

Static coefficient of friction, and therefore general floor polish safety, is an enormous area of debate and conflicting information in the industry. This section presents papers that clearly define what exactly coefficient of friction (COF) is, how it is to be measured correctly in accordance with ASTM standards, and the pro's and con's associated with other means by which it is claimed that COF can be quantitatively measured. There is a general review of polish maintenance techniques and a critical paper encompassing the relationship between the static coefficient of friction of newly applied floor polish and that same finish, which has been subjected to various industry standard maintenance techniques over time.

The final group of papers discusses floor polishes in terms of the legal aspects associated with a slip incidence and also regulatory issues that impact these coatings. The information provided here gives facility managers the tools to help be proactive in preventing slip incidents and some thoughts on the proper manner in which to respond if such an event should occur. Through presentation of an actual case history, one author provides a meticulous investigative outline that is applicable to those involved in the discovery phase of a slip claim. Lastly, the publication closes with a review of those regulatory issues that currently effect floor polishes.

Significant and pertinent information is presented here in relationship to floor polishes and the body of knowledge that is currently available. Although much of this work will remain relevant as technology in the field of polymer chemistry progresses, the information found here must be viewed as a foundation from which ASTM Committee D21 will need to build as advances in this scope of interest come forth.

William J. Schalitz
Spartan Chemical Company, Inc.
Maumee, OH
Symposium Chairman and Editor

BASIC FORMULATION CHEMISTRIES

Joseph M. Owens[1]

It's What's on the Inside that Counts – The Chemistry of Floor Polishes

REFERENCE: Owens, J. M., **"It's What's on the Inside that Counts – The Chemistry of Floor Polishes,"** *Technology of Floor Maintenance and Current Trends, ASTM STP 1448,* W. J. Schalitz, Ed., ASTM International, West Conshohocken, PA, 2004.

ABSTRACT: The chemicals that are typically blended to produce a commercial floor polish are listed and discussed with an emphasis on the properties that each brings to the final formulation and polish film. The distribution and function of the components is followed through polish application and film formation.

KEYWORDS: polishes, polish properties, polymers, monomers, metal crosslinking, waxes, dispersion particles, solvent classifications, coalescent, plasticizer, defoamer, wetting agent, leveling agent, stabilizer, biocide, slip resistance, gloss, durability, water resistance, detergent resistance, removability, scuff resistance, soil resistance, black mark resistance, film formation.

Introduction

Floor polishes represent a body of technology that has evolved over about three thousand years to the present. Even if we do not dwell on its historic milestones, it is a very complex technology that has evolved under the influence of changing performance requirements and advances in 'what is possible' from technology improvements. In the context of this document I will only present the current state of the art of floor polish compositions and formulations. Because floor polishes also represent a commercial industry, which responds to local as well as global performance needs, there is no single definitive polish composition. Therefore, this presentation of floor polish compositions will be broad, and generic. Other floor finish formulations, such as sealers for wood, tile, ceramic, and mineral floors, and household (consumer) polishes will not be discussed directly. Though household polishes are somewhat related in terms of technology and composition, the performance requirements are so different that they are considered to be completely different entities. Sealers are not intended to provide easy chemical removability, and some are not subjected to direct pedestrian traffic, so these simpler formulations will not be covered in detail, though the pertinent differences in their formulating should be evident from the discussion of removable industrial and institutional polishes in this paper.

Table 1 presents an abbreviated list of the performance features provided, to varying extents, by industrial floor polishes.

[1] President, J. M. Owens, Inc., 128 Oxford Lane, North Wales, PA, 19454-4400.

TABLE 1–*Polish performance properties.*[a]

Gloss [P]	Gloss Retention [P]	Scuff Resistance [P]
Black Mark Resistance [P]	Soil Resistance [P]	Cleanability [P]
Hardness development [A]	Film Hardness [P]	Viscosity [P & A]
Abrasion Resistance [P]	Water Resistance [P & A]	Removability [P]
Alkaline Detergent Resistance [P]		Scratch Resistance [P]
Plasticizer Migration Resistance [P]		Slip Resistance [P & A]
Recoatability [P & A]	Film Clarity [A]	Dry Time [A]
Stain Resistance [P]	Wetting [A]	Leveling [A]
Wet Abrasion (Scrub) Resistance [P]		Powdering Resistance [P]
Film Color and Color Stability [P & A]	Disinfectant (Quat and Phenolic) Resistance [P]	
Foam Control [A]	Freeze / Thaw Stability [P & A]	Storage Stability [P & A]
Spray Buff Response [P]	High Speed Burnish Response [P & A]	

[a] Superscripts denote the class of ingredient that is primarily responsible for the property as a result of level and selection:
P = performance ingredients; A = Application ingredients; P & A = both.

All commercial polishes provide these performance features to varying extents, but there is no single polish formulation that excels in all of them. Many of the properties are mutually exclusive in that maximizing one depresses performance in another. One of the tasks of the formulator is to select the appropriate balance in these conflicting properties to provide the best balance for his customer's needs.

Polishes are liquid mixtures that require a high degree of technical sophistication and accuracy in their manufacture and design. However, a description of the manufacturing process and the multitude of standardized tests that are an integral part of the product design, qualification and performance testing, physical properties, and quality control testing, will be bypassed.

Though manufactured as liquids, polishes perform as extremely thin, solid films. This exposition will conclude with a description of the processes involved in transforming the liquid formulation into a solid film.

Performance Ingredients

The performance ingredients are those that determine the maximum performance of the polish formulation in terms of durability (scuff resistance, mark resistance, soil resistance). These are embodied in the polymer, wax, and alkali soluble resin. The performance ingredients also determine the maximum performance in the polish wet-test properties (detergent resistance, removability, wet abrasion resistance, water resistance). Other performance properties, such as gloss, are not absolutely fixed by the performance ingredients but are set in a range of performance. For instance, the polymer broadly establishes gloss performance. Gloss may be improved by the level and selection of some of application ingredients, but only at the expense of sacrificing some durability. Durability is set by the polymer selection, and the level and selection of other ingredients cannot enhance durability beyond the limits that are established by the chemical compositions and combination proportions of the performance ingredients.

(The term 'durability' is a composite of scuff resistance, black mark resistance, gloss retention, and soil resistance. Heavier emphasis is placed on soil and black mark resistance, since failings in these areas is more difficult and costly for the building manager to correct or repair.) In Table 1 the commonly evaluated performance properties of floor polishes are attributed to the classes of ingredients (performance or application) that are primarily responsible for determining the level of the polish performance property.

TABLE 2–*Generalized Institutional Polish Formulation*

Material	Amount (%)	Primary Function
Performance Ingredients	(solids)	
Acrylic (co-)Polymer	19	Durability, gloss, wet test
Poly(Olefin) polymer - Wax	2.3	Durability
Alkali Soluble Polymer	1.1	Removability, film formation
Application Ingredients	(active)	
Water	66.8	Viscosity control
Glycol ethers	5.1	Coalescent: film formation
Phthalate esters	1.5	Plasticizer: film formation
Fluorocarbon Surfactants	100 ppm	Wetting
Phosphate ester	1.5	Leveling & film formation
Silicon oil	100 ppm	Foam Control
Nonyl and Octyl phenol surfactants	0.5	Freeze/Thaw and storage stability
Iso-Thiazalone	5 ppm	Bacteria and fungal control
Alkyl and aryl esters	2	Perfume and odor masking
Organic dye	0.2	Formulation color

Polymer

The polymer constitutes the largest part of the solids in the polish formulation and thus, of all the ingredients, has the greatest impact on the performance of the polish film.

Polish polymers are relatively high molecular weight (1×10^5 to 1.5×10^6 g/mole) reaction products of the random free radical polymerization of acrylic co-monomers. The polymerization is carried out by charging monomer, and small amounts of initiator and surfactant to a relatively large amount of water. The initiator is cleaved by either heat or a redox reaction to produce free radicals with a highly reactive, unpaired electron. These react with intercepted monomer molecules to form another free radical moiety, called an activated monomer. When the free radical activated monomer encounters another monomer molecule, they react, forming a new carbon-carbon bond to make a free radical dimer. Repeated monomer encounters and reactions lead to the formation of additional carbon-carbon bonds and trimers, quatramers, pentamers, hexamers, etc. Eventually the reaction outpaces the classical education of chemists, and they simply denote the products as polymers (poly = many). This random free radical polymerization process is shown pictorially in Figure 1.

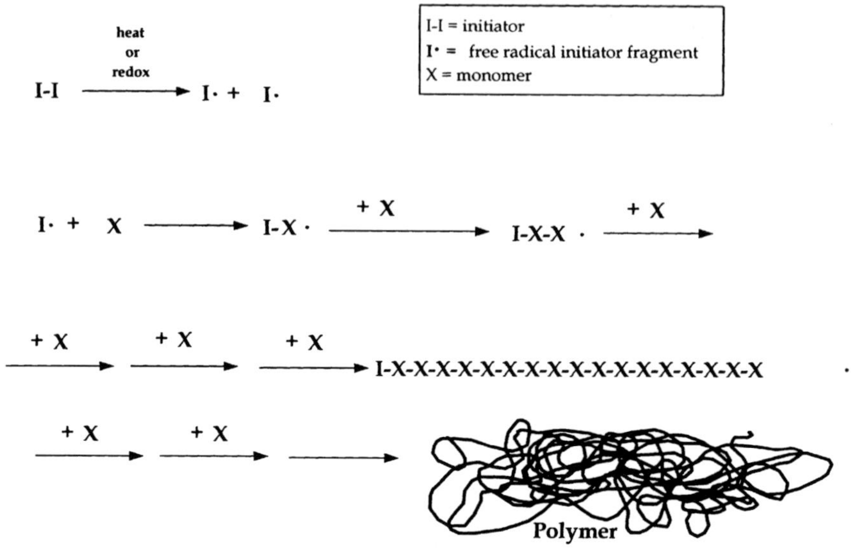

FIGURE 1–*Free radical polymerization*

As the molecular weight of the growing polymer increases, the physical and chemical properties of the molecule gradually change. The most obvious physical property change is solubility. For instance, we are all familiar with the high solubility in water of common sugar, a dimer of glucose. This high solubility is very different from the insolubility of cellulose or wood, which is chemically identical to sugar but is a high molecular weight polymer of glucose. For floor finish polymers, the most important physical property change that occurs as the molecular weight is increased is the ability to resist damage by pedestrian traffic. Figure 2 shows the variation of durability and gloss of polymers of differing molecular weights but of similar compositions and formulated with the same choices and levels of other formulation ingredients.

As each molecule of monomer is added to the polymer chain, a carbon-carbon double bond is lost and a new carbon-carbon single bond is formed. About 145 kcal of heat are released for each mole of monomers (60 to 200 g) added to the polymer. In order to control this tremendous amount of heat, the polymerization is carried out in water. Water is an ideal reaction medium because it has a very high heat capacity, high thermal conductivity, high boiling point, and it is cheap. The initial step in the polymerization is dispersing the liquid monomers in water. This mixture is an emulsion because it is a liquid dispersed in a liquid, but because the polymerization reaction yields a solid polymer dispersed in water, the final product is a dispersion. The terms emulsion and dispersion are often used interchangeably in the polish industry to describe the polymer (and sometimes the wax), though dispersion – material of one phase dispersed in material of a different phase – is the technically correct one. Water thus becomes part of the polish formulation. It serves other functions in the formulation, as will be discussed later, but its purpose at this point is assisting in the manufacture of the polymer.

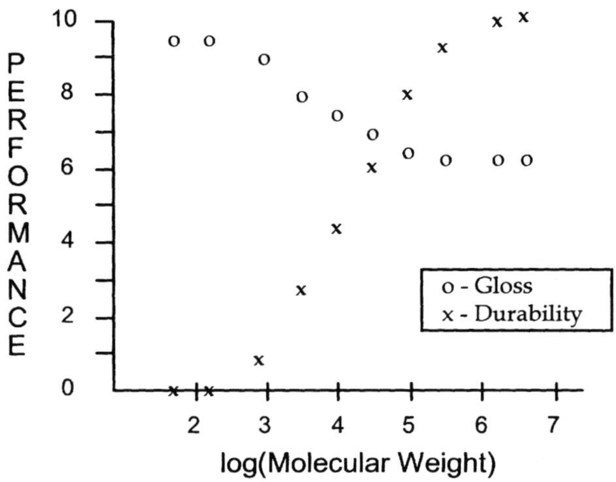

FIGURE 2–*Polish performance as a function of molecular weight.*

To ensure that the dispersion is stable toward long-term storage, freeze/thaw cycles, and mechanical agitation, small amounts of surfactant (called primary emulsifier) are added to the polymerization vessel or the monomer mixture (or both). The amount and choice of primary emulsifier, as well as the ratio of total emulsifier to monomers, determines the size of the dispersion particles produced. It has been found that smaller dispersion particles provide higher gloss in the formulated polish, and smaller dispersion particles require lower levels of solvent to attain good film formation (see below). Most polish polymer dispersions have a final particle diameter in the range of 80 to 120 nm.

Though smaller particle size dispersions are possible, they are not made on a commercial scale (except for household polishes) because the higher level of primary emulsifier needed will impact negatively on the polish properties of water resistance, wet abrasion resistance, and recoatability.

The monomers most commonly used in the manufacture of the polish polymer are listed in Table 3, along with their chemical structures and glass transition temperature (Tg). Other monomers, such as polymerizable surfactants or non-ionogenic hydrophilic hydroxyethylmethacrylate, are sometimes also included in the polymer to provide special or unique performance properties, but they are not common because of their cost or the manufacturing difficulties they present. Ethylene and propylene are gaseous at room temperature and so are not used in the manufacture of polish polymers, but they are the main building blocks for synthetic waxes (see below).

The first criterion for selecting monomers is the Tg of the resulting polymer because this will broadly determine the polymer minimum filming temperature (MFT), which, in turn, establishes the amount of solvent that will be required for film formation. The Tg and MFT can be predictably manipulated by combining monomers of different Tg's. MFT's can be further manipulated by adjusting the proportion of monomers that are hydrophilic so that the polymer is more readily solvated by water.

The second criterion for monomer selection is to introduce functionality to the polymer that will provide some specific performance properties. For instance, acid functionality is required for metal crosslinking (see below) and styrene monomer provides high initial gloss.

Finally, monomer selection for polish polymers is governed by concerns for the overall hydrophilicity of the polish film, because this general physical parameter has a significant influence on durability in traffic as well as the amount of solvents required for film formation.

TABLE 3–*Monomers commonly used in Floor Polish Polymers*

Acrylic Monomers		Tg (°C)
$H_2C{=}C(H)CO_2H$	Acrylic Acid (AA)	150°
$H_2C{=}C(CH_3)CO_2H$	Methacrylic Acid (MAA)	180°
$H_2C{=}C(CH_3)CO_2CH_3$	Methyl Methacrylate (MMA)	105°
$H_2C{=}C(H)CO_2C_2H_5$	Ethyl Acrylate (EA)	-22°
$H_2C{=}C(H)\ CO_2C_4H_9$	Butyl Acrylate (BA)	-56°
$H_2C{=}C(H)CN$	Acrylonitrile (AN)	150°
Other Vinyl Monomers		
$H_2C{=}CH_2$	Ethylene (vinyl) (Et)	-40°
$H_2C{=}C(H)CH_3$	Propylene (Pro)	-20°
HC-CO ‖ O HC-CO	Maleic Anhydride (Manh)	(high, undefined)
$H_2C{=}C(H)C_6H_5$	Styrene (vinyl benzene) (St)	100°

The phenomenon of a glass transition is similar to the melting point transition (solid to liquid) of a pure solid. The polymer is not a pure solid because the random character of the polymerization and radical termination processes give it a distribution of molecular weights, so for polymers the transition is from a glassy, hard state to a flexible, pliable, plastic state. The Tg is the mid-point of a range of temperatures over which the physical changes take place. At temperatures below their glass transition, all polymers are essentially equally hard. At temperatures above their Tg the amorphous polymers are soft, flexible, malleable, and ductile, and the magnitude of these properties is dependent to a great extent only on the temperature differential between the Tg and ambient temperature. Toughness, which is more critical than hardness for polish durability, is determined by the polymer molecular weight and the extent of chain entanglements, and crosslink density.

Of the monomers that provide functionality to the polymer, the acidic monomers are the most important, because they provide the polymeric sites for metal crosslinking (see below). The next most common monomer used to provide unique performance is Styrene. All other variables being held constant, Styrene provides uniquely high gloss to a polish film, though at the cost of significant reductions in removability, mark resistance, and soil resistance and minor negative impact on scuff resistance, plasticizer migration resistance, and film color stability. Styrene also has a positive, though minor influence on detergent, wet scrub, and water resistance, and resistance to quaternary amine disinfectants. The minor effects, both positive and negative, can be readily duplicated or offset by changes in the polymer composition or formulation, but the major effects on gloss and mark resistance are not otherwise available or correctable. The effect of polymeric styrene monomer content (stated as the percentage of total monomer weight) on polish performance is shown in Figure 3.

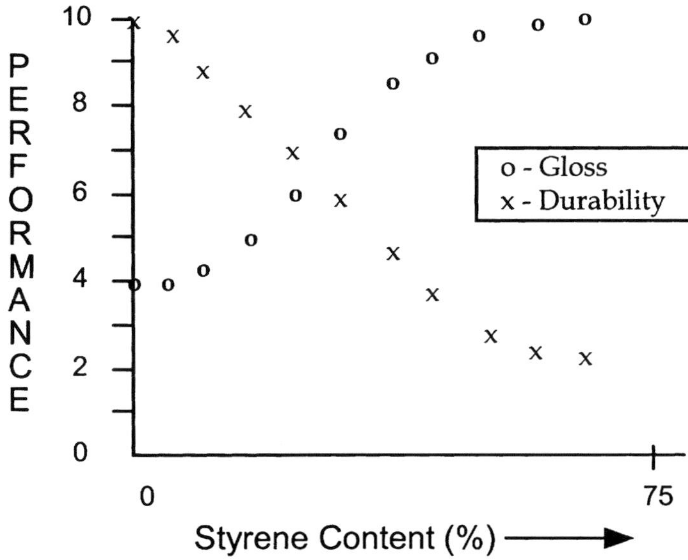

FIGURE 3–*The Styrene effect.*

The physical and/or chemical basis for the styrene effect on gloss is not fully understood, even though it has been the subject of extensive investigation for at least 25 years. The fruit of this futile investigation is a lot of information that demonstrates that the hypothesized causes are *not* responsible for the styrene effect. The elevated gloss of styrene-containing polymers is *not* due to the aromatic character of the polymer, residual monomer or low molecular weight polymer fragments, the refractive index of the polymer, polymer density, different particle packing characteristics during film formation, nor a number of other possibilities that were considered and tested. These erroneous theories will not be discussed here because, in retrospect, many now appear so far-fetched as to be embarrassing.

On the other hand, the styrene effect on polish durability is well understood. Of the standard monomers used in polish polymers, only styrene is totally non-polar and non-polarizable. It is very hydrophobic. The presence of styrene in the polymer backbone makes the polish film very hydrophobic and thus very compatible with the hydrophobic materials of shoe heels and soles (usually constructed of Styrene/Butadiene rubber). This compatibility makes it easier for the black rubber to be driven into the film, carrying with it the dark pigment of the shoe material, leaving a black mark in (rather than on) the polish film. The hydrophobicity of styrene-containing polymers also increases the film's compatibility with hydrophobic soils and oils (poorer soil resistance) and hydrophobic plasticizers that can migrate from the tile into the polish film (poorer plasticizer migration resistance). At the same time, the hydrophobicity of styrene decreases the film's compatibility with polar materials such as water (improved water and wet scrub resistance), aqueous detergents (improved detergent resistance, offset by more difficult removability), and quaternary amine disinfectants (improved resistance to "quat burn"). The reduced scuff resistance of highly styrene-containing polish polymer films is due to the propensity of the hydrophobic aromatic rings to associate and cluster in the (relatively) hydrophilic matrix of the remainder of the polymer, forming very hydrophobic pockets in the film that retain plasticizer in high concentrations and keep the film soft.

For some polymer manufacturers, styrene is a less expensive monomer than the acrylic monomer alternatives, so high styrene content polymers are preferred by them to reduce their raw materials cost.

With the exception of high initial gloss and low manufacturing costs, all of the positive effects of styrene polymers in floor finishes can be attained by other polymerization or formulating strategies. The negative effects of styrene polymers in floor finishes are not susceptible to correction by manipulating of other variables.

Metal Crosslinked Polymers

As noted above, polish film toughness and durability is derived from the polymer high molecular weight and crosslink density. Crosslinking during the polymerization process would make subsequent film formation impossible or extremely difficult, so the crosslinking reaction must take place after the polish has been applied to the floor. In floor polishes, this latent, post-application crosslinking is provided by metal crosslinking.

Metal crosslinking technology is a workhorse of the industry, being the basis for the majority of commercial janitorial floor polish formulations. Initially developed to provide a balance of detergent resistance and removability hitherto unavailable, the technology has evolved to allow polymer composition modifications that give much higher durability than was possible with low functionality polymers. This evolution in durability performance was possible because the crosslinking not only increased the effective molecular weight of the polymer, it also allowed the use of higher levels of acid functionality in the polish polymer without compromising water and detergent resistance. Because the metal reacts with the pendant acid functionality rendering it inert to alkaline materials, high levels of acid are possible without compromising resistance to alkaline detergents. The increase in acid content also increases the film hydrophilicity or polarity, providing a further boost in durability (the opposite of the styrene effect on durability noted above). Metal crosslinking technology is *not* used in formulating seals, which have no need for augmented removability, and household polishes, which have no need for augmented detergent resistance or durability.

Though most of the transition metals are suitably reactive to serve as the metal in metal crosslinked polymers, considerations of cost, toxicity, and color (of the liquid polymer dispersion, the polish formulation, and the finish film), limits the choice to Zinc for all commercial metal crosslinked polish polymers.

Zinc metal crosslinking makes the acid functionality in a polish film inert to alkali by reacting to form a coordinate covalent ligand structure with one metal ion and two carboxylic acids, as shown in Figure 4. Because the acid functionality is randomly distributed throughout the polymer backbone, and the polymer chains in the film are intimately entangled (not shown, for simplicity), the crosslinking reaction produces crosslinks that locks the entanglements while increasing the molecular weight of the already high molecular weight polymer. High molecular weight in polish polymer films provides high levels of film toughness and abrasion resistance, as well as improved water and detergent resistance.

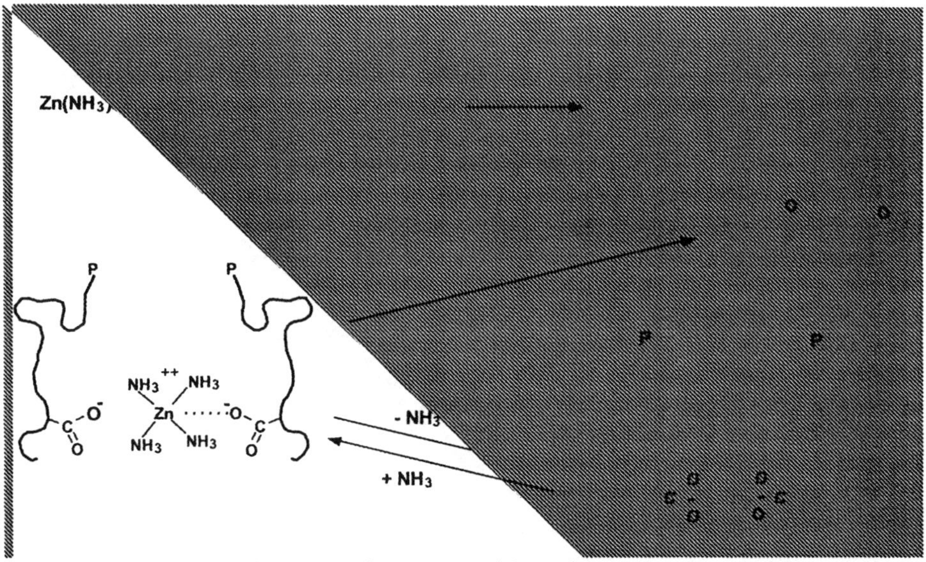

FIGURE 4–*Mechanism and Structure of the polymer/Zinc crosslinking complex.*

The ligand structure of the final polymer - Zinc crosslink complex (the lower right structure in Figure 4) is very sensitive to amines. The lone pair of electrons of the amine nitrogens attacks the structure by displacing carboxylate oxygens from the Zinc complex, breaking the complex into its constituent acid and metal ion (now with amines in the metal complex ligand structure). The liberated acid functionality is now free to be fully solvated by water to bring about the swelling, disintegration, and removal of the polish film. Because of its high molecular weight, the polymer cannot dissolve, but it has the toughness of freshly prepared jelly and can be easily abraded from the floor by mechanical action.

The formation of the crosslinked polymer complex during film formation and the disruption of the complex during polish removal follow the same reaction pathways, but the conditions or reagents present shift the direction of the equilibria. Zinc crosslinking gives the polymer a balance of resistant to alkaline detergent scrubs while making it sensitive toward amines and easing removability.

Other amines, such as alkyl amines, may be used in place of ammonia, because the reaction criteria is that the amine have a readily available electron pair to be donated into the low-lying d-orbital of the metal ligand site. Though hyperconjugation from the alkyl group makes these alternative amines better ligands, steric hindrance makes problematic the use of tertiary amines with fewer than two methyl substituents. The alkyl amines are also very much more expensive (particularly on a molar basis) than ammonia, they have a strong (fishy) odor, and they have lower volatility than ammonia. The lower volatility means that the final step in the crosslinking process is delayed and the film develops its toughness only after an appreciable amount of time. Any free alkyl amine in the film also serves as a solvent, softening the film and further detracting from early durability.

The pH of metal crosslinked polymers and their polish formulations is kept high with ammonia. This excess ammonia serves to shift the equilibria of Figure 4 so that the ligand sites of the metal are fully occupied with amines, preventing premature crosslinking (before the polish is applied to the floor) of the polymer. The excess amine also ensures that all of the acid functionality available to the aqueous phase is neutralized as ammonium salts. The polymeric carboxylate ions give the dispersion particle a strong negative charge, ensuring the stability of the dispersion. However, the excess ammonia also gives the formulation a strong odor that is sometimes found objectionable.

Some modern polish polymers use a crosslinking technology that does not require the intermediate amine ligands for the metal. The final crosslink structure is identical to the complex shown in the lower right of Figure 4, however, and removability is obtained by way of the standard mechanism and process. The lower amine content of these polymers means that polishes can be formulated with almost no ammonia odor. This technology, called "metal oxide crosslinking" is still under patents, so it is not yet universally used in industrial polishes.

Just as formulating a polish is both an art and science, the design of a floor polish polymer is based on an extensive amount of science leavened with art and experience. Both arts are full of compromises and fine balancing of the performance properties made possible by the science.

Wax

Most commercial polishes do not contain wax. Waxes are specific chemical compounds, derived from natural animal or vegetable sources. Waxes gave their name to polish formulations when the formulations consisted of just solvent and waxes or water and waxes. In some places the name 'wax' still persists for polish formulations in spite of the absence of wax in modern floor finishes. (Polish or finish is the correct term.)

The natural waxes of earlier polishes have been displaced with low molecular weight poly(ethylene) and poly(propylene) polymers. Olefin polymers have displaced natural waxes because the synthetics are cheaper, lower in color, more chemically and mechanically stable, and provide better control of slip resistance. The olefin polymers are often called 'waxes' in the industry, because the synthetics serve the same purpose in the formulation as the original natural waxes. Waxes are the second of the three performance ingredients in polishes.

Commercial olefin-based synthetic waxes offer a relatively narrow range of molecular weights, compositions (propylene or ethylene), and functionality (co-polymerized acrylic or maleic acids, or post-oxidized after polymerization), and degree of branching (a function of the polymerization processing variables), but they provide a sufficient range of performance in polishes that the selection and possible blending of 'waxes' in a formulation is a significant aspect of the polish formulator's art.

Olefin polymers are manufactured and sold as solid materials that must be melted and dispersed in water before they can be added to the aqueous polish formulation. Because they are largely non-functional and very hydrophobic, the polymers are difficult to emulsify. Emulsification requires fairly heavy charges of surfactants, and these surfactants contribute significantly to the solids of the wax dispersion product as well as have an impact on the performance of the polish film. Because the emulsifying surfactant is an essential and integral part of the wax, it is considered as part of the wax and its effect on polish performance is grouped into that of the wax dispersion.

The olefin polymers are capable of forming crystalline microregions through laminar association of the regular hydrocarbon chains. This gives them an elevated Tg and additional toughness, beyond what would be expected of their low molecular weight. Much of this micro-crystalinity is lost when the wax is emulsified.

Because they are difficult to emulsify, wax dispersions typically have particle diameters in the range of 150 to 250 nm, much larger than the particle size of the polish polymer dispersion. This larger particle volume for the wax dispersion means that an equal weight of polymer solids will comprise about eight times as many particles.

The selection and amount of waxes in a polish formulation impact the film's performance in scuff resistance, stain and soil resistance, color, and slip resistance. The surfactants that are charged with the wax, from the emulsification process, impact negatively on the film's performance in recoatability, and water, soil, scuff, and detergent resistance, and positively on the properties of film formation, leveling and wetting, and removability. Though the wax has no effect on the gloss of a polish film, it has a significant effect on the gloss build that is possible from spray buffing and high speed burnishing.

The effect of the amount of wax charged to a formulation on scuff resistance performance must be empirically determined for each formulation and combination of waxes. As a formulation is developed or modified, a plot of performance *vs.* concentration, such as shown in Figure 5, must be constructed by the formulator.

The maximum in a performance plot such as Figure 5 will shift to higher or lower wax content levels as the character of the polymer is changed and as some application ingredients, particularly the plasticizing solvent level and the amount of free surfactants, are changed. The initial slope leading to the maximum and the trailing slop of the plot are a function of the hardness and slipperiness of the wax. The polish formulator's art is in recognizing that the optimum wax content for the formulation may not be at the maximum of the scuff resistance/wax content curve but at some other point that provides other desired properties. For instance, because waxes are softer than polymer, shifting to a higher than optimum wax content will provide a softer finish, which will have improved high speed burnish response. The less-than-optimum scuff resistance is not begrudged because film scuffing is readily repaired by high speed maintenance.

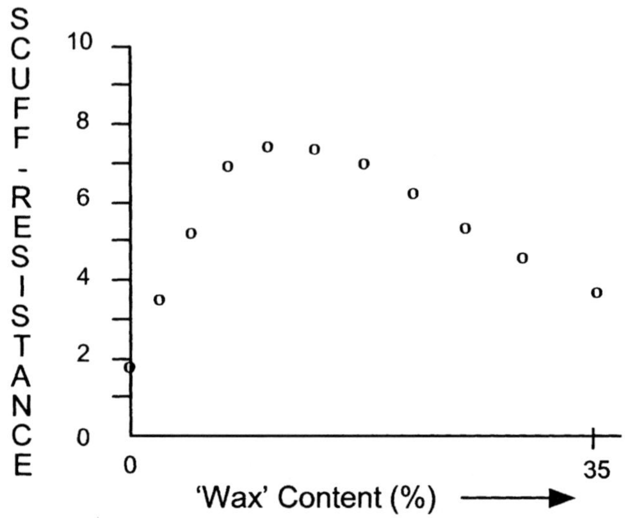

FIGURE 5–*Polish scuff resistance as a function of wax content.*

Alkali Soluble Resin

The third of the performance ingredients are alkali soluble resins (ASRs). ASRs are low molecular weight, very high acid materials that, as their name implies, dissolve in aqueous alkaline solutions. The materials are either derived from pine rosins (called modified rosin esters) or synthetic resins made from the same array of monomers (see Table 3) as the polymer. Another class of synthetic resin is made from styrene and maleic anhydride, but these have been largely replaced by synthetic ASRs, because the styrene effect on gloss does not exist for low molecular weight (< 100 000 g/mole) polymers so there is no gloss advantage to high styrene content in ASRs, and high styrene content inhibits solubility so that the molecular weight of the ASR must be made extremely low to retain solubility in aqueous alkali.

Modified rosin ester ASRs are generally lower in cost than synthetic ASRs, but they suffer from wide variations in quality (depending on the properties and location of the trees from which they are derived), and they are dark brown in color. This inherent color is intensified by age and heating, and heating is required to make the resin into an ammoniacal solution before it is added to the polish formulation.

As neutralized (ammonium salt) solution polymers, all ASRs form a film with the simple evaporation of ammonia and water, in the same way that a sugar solution forms a film with the evaporation of water. As solution polymers, they do not display a

MFT, and so the polymers can have a very high Tg with no attendant film formation problems. The only limitation to the Tg of ASRs is the fracturing that might occur in the very brittle film as it shrinks from water loss during the later stages of the drying process. Because of their low molecular weight, the films are very hard and brittle, but they are not tough.

The molecular weight of ASRs used in polishes is limited by concerns for solution viscosity, because as molecular weight increases the solution viscosity drastically increases exponentially (see Figures 6 and 7). The high acid level in the ASR's, coupled with the ammonia or other bases in the polish formulation, means that they are solutions and cannot exist as stable dispersions (though some synthetic ASRs are manufactured and sold as dispersions, these are converted to solutions when charged to the high pH polish formulation).

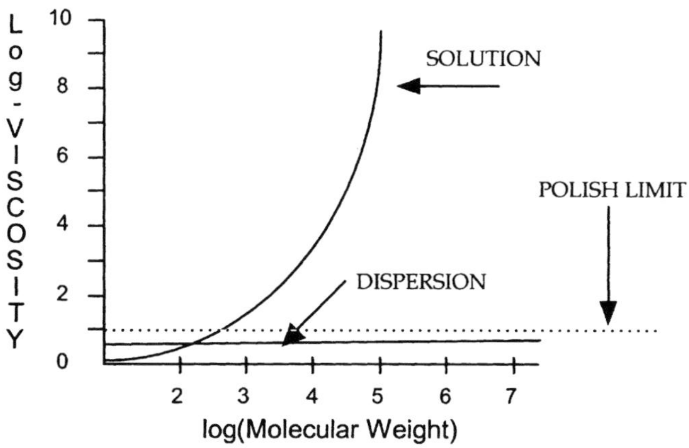

FIGURE 6–*Solution viscosity as a function of molecular weight, at fixed solids.*

Though the loss of ammonia makes a film of ASR less soluble, it still retains a high degree of water sensitivity and an extremely high degree of alkali sensitivity. These properties are imparted to the polish film when an ASR is included in a formulation, and the extent of these sensitivities is in proportion to the amount of ASR incorporated. ASRs also impart slightly improved gloss and leveling performance, and better removability. Because these latter properties can be derived from other formulation ingredients, the negative effects on film toughness and water and detergent resistance are important limitations on the amount of ASR incorporated in the finish formulation. For this reason ASRs are kept to a minimal proportion of the polish solids in industrial polishes, though they are the major formulation ingredient in household polishes.

ASRs are added to a polish formulation to improve the efficiency of the coalescing solvent (see below) in promoting film formation of high molecular weight polish polymer dispersion. The presence of an ASR in a formulation will allow as much as 10% lower coalescing solvent to be used without compromising film formation. Not only is the ASR usually the lowest cost of the three performance ingredients, the coalescent efficiency improvement can constitute a significant savings in the formulator's raw materials cost.

Because ASRs are low molecular weight solution materials they naturally enhance the gloss of the polish film (again, think of a sugar solution). If the polish formulation is maintained at constant solids, then the partial substitution of ASR solids for polymer solids means that there is less of the tough polymer present in the film, reducing the overall durability of the polish.

Water

Since all of the Performance Ingredients are added as aqueous dispersions or solutions, water is already present in the formulation. Water is considered to be an application ingredient, however, because *additional* water is added to the formulation to establish the viscosity of the final formulation.

Pure water has a viscosity of one centipoise (this is the definition of the centipoise unit). In order to flow out and level properly, the viscosity of the polish formulation must be less than about 10 cP. To put this value in perspective, if you were to add a teaspoon of sugar to a cup of coffee or tea, and then allow it to cool to room temperature, the viscosity would be 12 to 15 cP. The solution is not only unfit to drink, but it would be too viscous to be used as a polish.

The addition of water controls the viscosity of the formulation by reducing the solids content. The effect is shown in Figure 7. For most polish formulations the viscosity limit imposes an upper limit on solids of 25% to 30%. There is no lower limit to solids imposed by viscosity, but there is a practical lower limit of about 3% solids imposed by the need to form a coherent film capable of withstanding traffic.

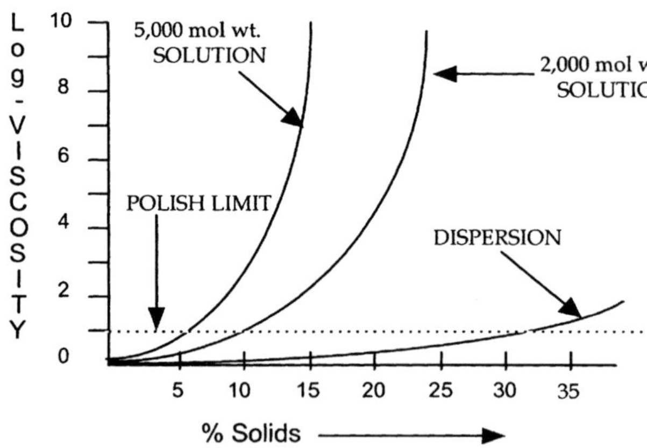

FIGURE 7–*Solution viscosity as a function of solids content.*

Between these upper and lower limits formulation solids determined by the desired formulation cost (polish solids have a *much* higher raw materials cost than does water) and required initial gloss. The gloss of a polish film is a function of the amount of material applied, as shown in Figure 8. Figure 8 also shows the initial gloss that comes about as a result of multicoating a polish (circles) incorporating a polymer with high levels of Styrene monomer, as was discussed earlier. Note that multicoating a polish formulated with a styrene-free (all acrylic) polymer (crosses) can provide the same gloss as a styrene-containing polish, though only when more material (coats) applied to the floor. Applying fewer coats of higher solids formulations can attain the same results. The application of fewer coats to attain a high film loading and high gloss reduces application time and cost.

The rate of gloss build with increased film deposition (the initial slope of the circles in Figure 8) is a function of the amount of styrene in the polish polymer, with higher styrene content producing a steeper slope. This is a manifestation of the 'styrene effect' on polish gloss that was discussed previously.

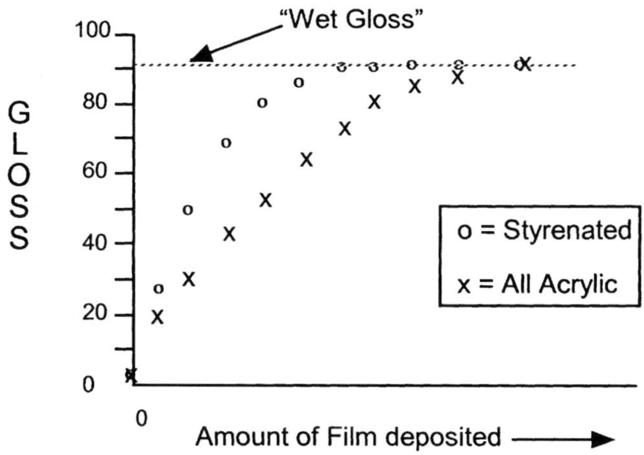

FIGURE 8–*Gloss as a function of film thickness*

The upper limit of gloss shown in Figure 8 is called "wet gloss" because the dried film has the same gloss as a floor wet with ponded water. When "wet gloss" is achieved it is not possible to tell if an applied polish is wet or dry. Once the "wet gloss" state is attained, the addition of more coats of finish does not further increase the gloss. Though the term 'wet gloss' is often applied to high speed burnishing, gloss from the mechanical process is not truly 'wet gloss' since it is possible to discern visually when water is spilled on a burnished floor by noting the difference in the gloss of the wet floor and dry film. "Wet gloss" is reproducibly attained only under the controlled conditions and careful application of polish in the laboratory and is not a realistic goal for field application. Because visually discerning the presence of a slip hazard such as a wet spill is important to pedestrian safety, true "wet gloss" is not a desirable condition for a floor finish in traffic.

Solvents

Polishes are formulated with two types of solvents to assist the high MFT, high molecular weight polymer dispersion to form a film. In other coatings applications, such as paints, these classes, plasticizers and coalescents, are distinguished simply by whether they remain in the dried film (plasticizer) or not (coalescent). This is not a distinction based merely on volatility, since the boiling points of many plasticizing and coalescing solvents are similar. In polish formulating these traditional distinctions are *generally* adhered to, but a more critical classification is made on the basis of where and how the solvent works in promoting film formation.

Coalescents

Polish coalescents are *often* relatively volatile, but they are *always* hydrophilic. Good polish coalescents have a degree of solubility in water as well as a degree of solubility in the polymer. When a hydrophilic coalescing solvent is added to an aqueous polish formulation, it partitions itself between the water phase and the polymer dispersion particles (and wax dispersion particles, but there are usually relatively few wax particles present, so they will be ignored in this discussion). Partitioning occurs at a fixed ratio or distribution of the solvent, depending on its compatibility with the polymer (which varies with the polymer composition). Solubility in both the water and polymer phases means that most of the coalescent is at an equilibrium position at the water/dispersion particle interface, with the remainder in the aqueous medium. The solubility of the coalescent in water, and water in the coalescent, keeps the coalescent from penetrating deeply into the interior of the (relatively) hydrophobic polymer particle. The affinity of the coalescent for the polymer causes the surface of the polymer particle to swell or expand in volume, allowing even deeper penetration of the coalescent into the particle. This is an essential part of the film formation process for the very thin films of floor finishes (see below). It is *not* possible to formulate a high Tg polish with good film formation performance without using a solvent that acts at the interface between water and the dispersion particles.

The coalescing solvents most commonly used in polish manufacture are in the families of alkoxyethanols and alkoxypropanols (mono-alkyl glycol ethers). Part of the art of formulating is in the selection of the coalescing solvent or solvent blend that maximizes solvating power for the polymer while not thwarting the solubility of the solvent in water. Obviously, the optimum blend of coalescing solvents changes with the chemical composition of the polymer. Solvents that do not maximally soften the polymer / water interface must be used at higher levels, which results in higher cost, prolonged dry times, and degraded polish recoatability.

Plasticizers

Polish plasticizers are *often* non-volatile (at least relative to coalescing solvents), but they are *always* essentially insoluble in water. Plasticizers are very soluble in the polymer phase of the formulation. Because of their hydrophobicity, plasticizers in a formulation equilibrate to reside in the interior of the polymer particles, as far from the water phase as they can get.

The plasticizers used in polish formulating are usually selected from the families of symmetrical phthalate esters, di-esters of highly branched alkane diols, or symmetrical phosphate esters. The alkaline aqueous medium of a polish will hydrolyze esters, and for this reason esters cannot be used as coalescents, but plasticizers quickly move to the interior of the particles and away from water, so ester hydrolysis is not a problem for plasticizers.

Polishes can be formulated with no plasticizing solvent, though this usually means that the amount of coalescing solvent must be increased significantly. In practice, plasticizers are present in almost all formulations, not only as a cost-control measure, but the leveling agent commonly used (see below) is also acts as a plasticizing solvent. Excessive amounts of plasticizer increase the formulation cost and, usually being of low volatility, remain in the dried film keeping the film soft and compromising durability. Non-volatile plasticizers also contribute to the formulation solids (an important benefit for specification polishes).

When properly formulated, a polish polymer dispersion particle will contain both coalescing solvent at and near the particle/water interface, and plasticizing solvent in the interior of the particle. If we start at the center of a formulated particle and move out along its radius, there is a gradient of increasing concentration of coalescent that goes from zero at the core to a maximum at the particle surface. Along this same axis there is a decreasing concentration gradient of plasticizer that is a maximum at the core and zero at the particle/water interface. These gradients are depicted schematically in Figure 9.

The combination of both plasticizers and coalescents in polish formulations means that the particle is fully solvated throughout its volume with the minimum amount of solvents. This reduces formulation costs without compromising film formation.

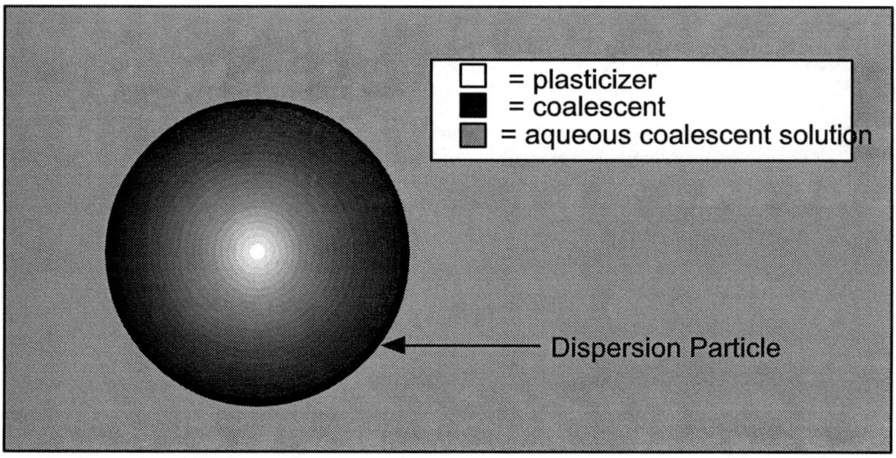

FIGURE 9–*Idealized depiction of solvent distribution in polish dispersion particle.*

Wetting Agent

Wetting performance is the ability of an aqueous polish to wet out over any surface that it may be expected to coat. These range from relatively polar or polarizable surfaces, such as acrylic and modified acrylic polish and sealer films to relatively hydrophobic surfaces, such as uncoated solid vinyl and vinyl composition tiles. To wet out properly, the formulation must have a surface tension that is as close as possible to the surface energy of the substrate. However, the surface energy of the substrate is even less predictable than the composition of the substrate, so optimum wetting is almost never achieved, and we must be satisfied with the compromise of "good enough" wetting over the more commonly encountered surfaces. A formulation surface tension in the range of 28 to 30 dynes/cm usually gives good wetting performance over most flooring surfaces. A formulation with surface tension significantly higher than the surface energy of the substrate will de-wet by spontaneously crawling back on itself (visualize water on waxed paper). A formulation with surface tension significantly lower than the surface energy of the substrate will de-wet by spontaneously "picture framing" or flowing to the edges of the surface where it dries to a slightly thicker, and therefore glossier, ridge.

The problem is not as dour (nor the compromises as severe) as the above may make it appear. Most flooring has become abraded and roughened with use, and this surface irregularity makes wetting much simpler because of capillary attraction to pores, fine scratches, and other surface irregularities. Some flooring types, such as vinyl composition tile and unglazed quarry tile, are inherently porous and irregular, and so these pose only a weak challenge for wetting. Bare terrazzo flooring (~40 dynes/cm is optimum for wetting over the concrete, but ~30 dynes/cm is optimum for wetting over most marble and marble chips) presents a unique problem of simultaneously wetting out over two surfaces of different surface energy. Terrazzo is best handled by sealing first so the polish must simply wet the acrylic seal surface.

The surface tension of the formulation is reduced from the 72 dynes/cm of pure water by surfactants from the wax dispersion, the presence of ASRs, and coalescing solvents in the aqueous phase. These usually bring the surface tension into the 40 –45 dynes/cm range. The final adjustment of formulation surface tension *could* be done by simply adding additional surfactants, similar to those used in making the wax dispersion, but this would require a heavy charge that would depress polish scuff, soil, water and detergent resistances and foam control. Instead, a special class of anionic fluorocarbon surfactants that is very efficient at reducing surface tension (and is a pretty poor surfactant otherwise) is used because only about 100 parts per million in the formulation is sufficient for good wetting performance.

Leveling Agent

Leveling performance of a polish is often mistaken for wetting performance, because inadequacy in either results in a streaky appearance (shallow ridges of uneven polish thickness that seen as streaks of high and low gloss). To distinguish between the two, the polish film must be observed while it is drying. If a wetting problem exists (a mismatch between the formulation surface tension and the substrate surface energy), the formulation will move to form the ridges while it is wet and mobile. If a leveling problem exists, the wet film will appear to be smooth and uniform until the formulation is in the later stages of drying, at which time the uneven gloss will appear. These leveling ridges appear to follow the high shear trail of the string mop or polish applicator, even when a coat of finish is evenly applied. To avoid leveling problems with a formulation, a specific, unique chemical leveling agent is charged to all polish formulations.

The leveling agent used in most commercial polishes is a symmetrical phosphate ester, *tris*-butoxyethylphosphate (TBEP). When charged to the formulation in the range of 1.0 to 2.0%, it almost magically provides leveling. TBEP is very expensive and it also functions as a very efficient plasticizing solvent in the formulation, so more than the minimum amount required for leveling is usually avoided.

The only published (patents) alternative to TBEP is a group of linear alcohol ethoxylate surfactants with a low level of ethoxylation (low HLB). These appear to work in the formulation in a different way than TBEP in providing leveling, though they are equally effective. The use of these surfactants suffers from some of the problems associated with excess surfactants mentioned previously. Though they do not work with every polish formulation, they are significantly less expensive than TBEP in those formulations in which they do work.

Polish leveling appears to be a function of the *dynamic* surface tension of the formulation. Although dynamic surface tension is not appreciably more difficult to measure than the static surface tension (the controlling variable in wetting performance), it has been very difficult to establish the appropriate target range of dynamic surface tension that provides good leveling performance. Since the connection of leveling performance to dynamic surface tension is theoretical and tenuous, formulators continue to rely on their art, experience, and TBEP to provide leveling performance.

Defoamer

Because of the various surface-active materials in the floor finish formulation, the common occurrence of polish agitation in the bucket and wringer, and the high shear of normal string mop polish application, all polish formulations have a tendency to foam. This foam is a problem only if it breaks so slowly (or the polish dries so rapidly) that residual foam marks are left in the dry film. To provide timely breaking of the inevitable foam from polish application, classic defoamer chemicals are used in the polish formulation. Antifoam chemical agents have been found to be ineffective in floor finishes. The most commonly used defoamers are based on silicone oils.

Polish defoamers must not only be totally insoluble in water but must actually repel water in order to break the water film that constitutes the wall of a foam bubble. These oils are almost impossible to emulsify in water so they have to be adsorbed onto a hydrophilic medium, such as silica gel, and this is then dispersed in water so that it can be added to the formulation.

As little as 125 ppm of active silicone oil is usually sufficient for good foam control. Excess defoamer can cause a problem of "fish eyes" in the dried film where the applied wet formulation is repelled or pushed away from a clot of defoamer particles so that the polish is not deposited evenly.

In Europe and some parts of the Pacific Rim the normal silicone oil defoamer is replaced by 2 to 8% fatty acid in the formulation. Fatty acids are slowly absorbed into the polymer and wax dispersion particles as non-volatile coalescents and the concentration of defoamer in the aqueous phase of the polish, where it must reside in order to be effective, decreases as the formulation ages. To compensate for the absorption during storage, high levels of fatty acid defoamer are used to provide long-term foam control. These much low cost defoamers do not cause "fish eye" problems when used at high concentrations, but they can cause problems of poor alkali resistance in the dried film as well as the problems associated with excess plasticizers. Unlike the silicone oils, fatty acid defoamers are biodegradable.

Stabilizer

After manufacture, formulations may pass through multiple freezing and thawing cycles while in transit or storage before they are applied to a floor. Depending on the quality of the wax dispersion used in the formulation, cycling through freezing and thawing may cause the wax dispersion to coagulate and separate as a thick cream or large, soft globules. A polish formulator can anticipate and avoid this problem by adding a small amount of surfactant stabilizer.

Polish freezing usually happens at, or very near, equilibrium conditions so that large ice crystals slowly grow in the formulation. These crystals disrupt and damage the wax emulsion particles, actually slicing them into smaller pieces, exposing surfaces of hydrophobic wax that were not coated with surfactant in the emulsification process. These new, unstabilized dispersion particles agglomerate and coagulate to cause viscosity instability (an increase above the 10 centipoise viscosity limit), graininess in the dried film, and creaming or sedimentation in the liquid polish. Stabilizing surfactants subvert these processes by forming micelles in the formulation that act as additional crystal nucleation sites. More nucleation sites means that more crystals will form, but each will be smaller and less likely to damage the wax emulsion particles. Depending on the surfactant choice, the micelles may also provide a reservoir of surfactant to coat newly exposed wax surface on particles that are disrupted or fragmented by freezing.

Freezing also disrupts the polymer dispersion particles, but the newly exposed surfaces contain acid functionality that is immediately neutralized by ammonia in the aqueous medium. This generates an anionic charge on the new fragment that stabilizes it against agglomeration in a manner identical to the stabilization given to the parent particles by the anionic surfactant used in the polymer's manufacture.

Anionic surfactants are most efficient at providing freeze/thaw stability, but they have a very strong tendency to foam. The non-ionic surfactants used to emulsify the wax also work, but only at concentrations that will otherwise impair polish performance properties. In practice, blends of anionic and nonionic surfactants are used at levels of less than 1.0 %. Most commercial formulations do not require stabilizing surfactants.

Biocide

Many formulations are warehoused or stored for long periods of time between manufacturing and application. During this time, particularly if the storage area is warm, fungus can grow in the formulation. The major nutrient for the fungus is the free surfactant in the formulation, so the first evidence of a problem of fungus growth is sedimentation, creaming, or viscosity increase in the formulation as the emulsifying surfactants that stabilize the wax are consumed. These thrills are quickly followed by very offensive odors.

Because the formulator cannot eliminate the nutrients for the fungus, a biocide, such as a chloro-*iso*-thiazalone, is used at 3 to 10 ppm. Except in North America, Japan, and parts of Europe, 500 ppm Formaldehyde remains the most commonly used polish biocide for industrial floor finishes.

Dyes

Dyes are used in some older formulations to hide the yellow/brown color of some ASRs and waxes. These date from the days when polishes were based on high levels of natural waxes and highly colored ASRs. Dyes serve to make the color imparted by the wax or ASR to the wet formulation, as well as the dried film, less obvious. The dyes are most often green or blue in color, and in the formulations they may also serve as a weak, non-volatile coalescing solvent (the dyes used in aqueous polishes are very water soluble). Because of their water solubility they can be washed out of the dried film by conventional floor maintenance procedures. This tendency, and their expense, explains why dyes are not commonly used in modern industrial and institutional polishes.

Dyes are not necessary in modern polishes, though in a few instances product color has become part of the product or corporate identity.

Perfumes

In Europe, Latin America, and parts of Asia, many industrial polishes are formulated with perfumes to mask the ammonia odor of the formulation or the ammonia that is evaporated from a drying polish film (only North Americans and northern Europeans have been successfully indoctrinated to the idea that ammonia is a 'clean' odor). The perfumes are usually specially formulated, proprietary blends of esters that may be part of the product or corporate identity, but they are usually quite expensive. Obviously, the lowest effective level is used in the formulation.

Because they are based on esters, perfumes have a tendency to be hydrolyzed in the basic aqueous medium of a polish. The resulting alcohols have very different odor qualities than the esters, which cause the odor of the formulation to change over time in storage. The acids generated from the hydrolysis of these esters are converted to the ammonium salts that are not volatile, so they make no contribution to the odor of the wet formulation.

Those perfumes that are more hydrophobic may slowly migrate out of the aqueous medium and into the more hydrophobic regions of the dispersion particles where hydrolysis can be avoided.

Polish Application and Film Formation

As was mentioned earlier, all of the application ingredients in a polish formulation are present to ensure that the performance ingredients will properly form a film on application. Poor or incomplete film formation is often the root cause of many polish performance problems, such as low gloss, poor water resistance, poor detergent resistance, powdering, and poor durability. Because so much of the floor finish formulating art is directed toward this essential process, we will now examine film formation in some detail.

Application

Polishes are conventionally applied with a string mop or applicator at a spread rate of about 2000 square feet per gallon. This "natural" spread rate is determined by the chemical nature of the fibers on the application instrument and the surface tension and viscosity of the formulation. If the polish formulation contains 15% solids, this application rate will result in a dried film thickness of about 2.5μm (1/10 000 inch, or one tenth of a mil). To put this in perspective, a single coat of paint is 40 times as thick as a single coat of dried polish film.

The application of thicker or thinner coats of polish than the "natural" loading (attempted to enhance or mitigate polish gloss) requires that the custodian make special mental and physical efforts in applying polish. Because this slows the application process, and thus increases the application labor costs, when heavier or lighter coats are desired it is usually easier to use a formulation with adjusted solids.

Relative Humidity

Polish film formation is highly dependent upon ambient application conditions. Most commercial formulations can be applied satisfactorily in an ambient humidity range from 15% to 85% relative humidity. A relative humidity range of 40% to 65% is ideal. At relative humidity higher than about 85% the equilibration of water back into the drying film from the air causes the polish to take too long to dry.

As shown in Figure 10, as soon as the wet formulation is spread over the very large surface area of the substrate, the loss of water begins, as does the loss of the ammonia (from the ASR and the latent Zinc crosslinker), and also the slow loss of volatile coalescents. (Because of their affinity for water, coalescents are lost very slowly until most of the water in the drying film has evaporated.) The lengths of the arrows in Figure 10 are roughly proportional to the rates of loss or gain, but only the water gain from the ambient air is dependent upon relative humidity. Polishes are formulated with a presumption that they will be applied within the "normal" range of relative humidity, so some return of water from the atmosphere is expected.

When the ambient air is quiescent, the air immediately above the wet polish quickly reaches 100% relative humidity. At this point the rate of water gain to the film from the air is equal to the rate of water loss from the film. Simply moving or stirring the air mass immediately above the polish will alleviate this situation, with the added benefit that moving air absorbs moisture more readily than does quiescent air. This is why floor fans are so effective at reducing polish dry time, particularly for polish applications at high humidity.

At very low relative humidity the rate of water gain to the film from the ambient air is very low, and the polish will tend to dry too fast. While it is not common to worry that the dry time of a polish is too fast, as we will see there are a number of chemical and physical processes that must be completed during film formation and these require a finite amount of time. If insufficient time is available because of too fast drying, the film will not display all of the properties that are designed into it.

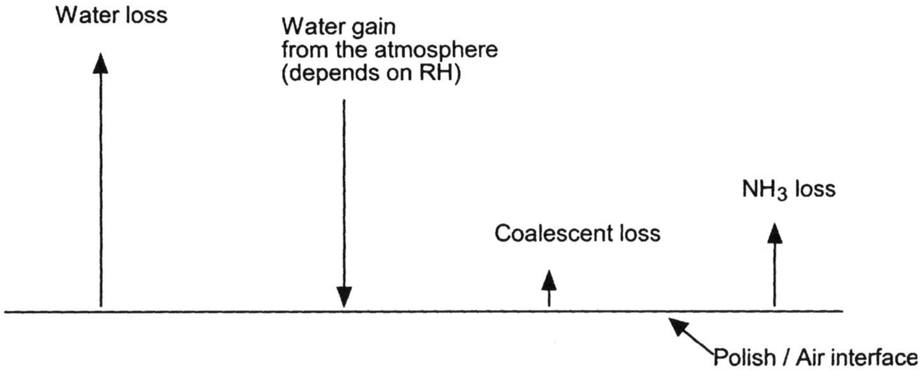

FIGURE 10–*Initial equilibrations with ambient air during polish drying.*

Floor Temperature

The other major ambient application variable in polish application is the floor temperature. Polish formulations are typically designed to form a good film on floors which are as cold as 50°F (10°C). Theoretically, formulations can be made to form a film at temperatures as low as the freezing point of the aqueous medium (32°F, 0°C), but the high coalescent and plasticizer levels required would make such a formulation economically unsound, not to mention the compromise this would force on polish durability, recoatability, and dry time. In most applications polish film formation at floor temperatures below 50°F is not necessary because floors are only very rarely colder than 55°F. Entryways in winter and open front freezer and dairy cases are areas where this could pose a problem.

The controlling ambient variable is the *floor* temperature, rather than the ambient air temperature because the high thermal conductivity of water and the huge thermal mass of the floor makes the formulation immediately equilibrate to the same temperature as the floor. Higher or lower air temperatures will change the relative humidity of the ambient air (see above), but otherwise air temperature has no effect on film formation until the floor equilibrates to the same temperature as the air, a slow process.

In Figure 10 the length of the upward arrows (rates of loss from the drying film) are dependent on the temperature of the formulation, and thus on the temperature of the floor.

The Drying Process

A depiction of the overall polish drying process is shown in Figure 11. The time axis is variable, depending on the floor temperature and ambient relative humidity.

As water evaporates from the drying formulation, the concentrations of ASR and coalescent in the remaining aqueous phase increases (the amount of these solutes remains constant but the volume of water decreases). Because the partitioning coefficient for the coalescent is a constant, determined by the solvent's relative affinity for water and polymer phases, the increase in coalescent concentration in the aqueous phase means that more of the solvent will migrate into the polymer particles.

The increase in ASR concentration with the loss of water means that the remaining aqueous phase is acquiring more ionic (salty) character. This reduces the solubility of the organic coalescent in the aqueous phase, by changing the portioning coefficient, and further increases the migration of solvent into the polymer. Because only that fraction of the coalescent that is at the air/water interface is available to evaporate, the rate of loss of coalescent remains very low.

When sufficient water has evaporated from the wet polish so that the formulation solids exceeds about 60%, the dispersion particles, which usually repel one another, are forced to come into contact with each other. The particles are highly solvated by plasticizer and coalescent, making them soft. Under the influence of very strong capillary forces, the soft particles distort, coalesce, and mesh together (See Figure 12). The high molecular weight polymer chains are now free to intermingle and entangle with polymer chains in adjacent particles, and eventually be crosslinked.

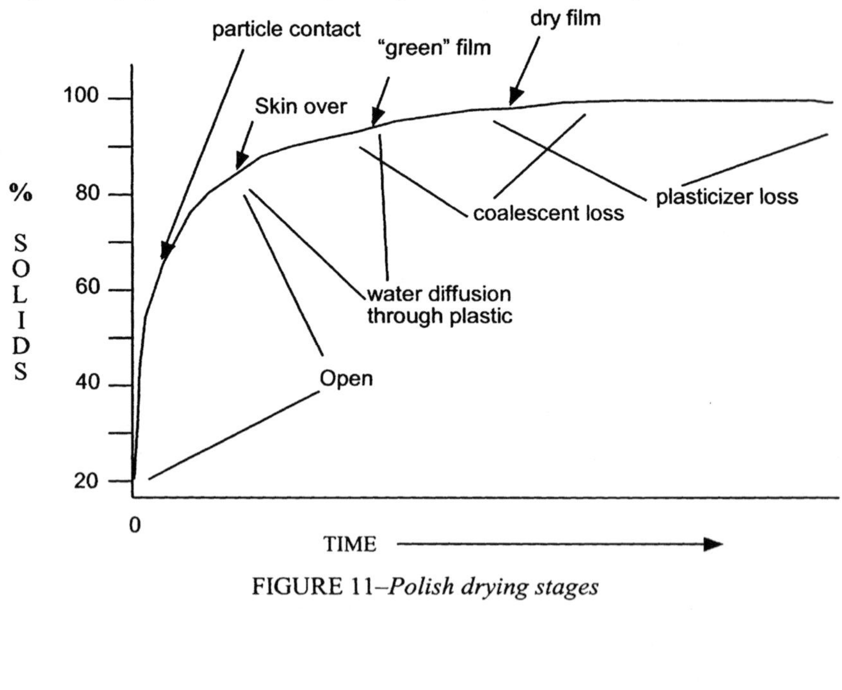

FIGURE 11–*Polish drying stages*

As additional water is lost and the continued coalescing of particles closes off all of the interstices or passages between particles and particle clusters, the polish surface becomes a plastic film. Because water loss occurs only at the polish/air interface, the film forms at the top surface of the polish over a polish mass that is still full of residual water. The underlying water-saturated polish is soft enough that touching the surface will leave a fingerprint or impression, but the surface is dry enough that wet polish is not transferred to the finger.

All subsequent evaporation must take place by water diffusing through the plastic surface film to get to the air interface, a much slower process. Once the water no longer exists as a separate phase in the nascent polish film, the coalescent, released of its affinity for water, will be free to evaporate.

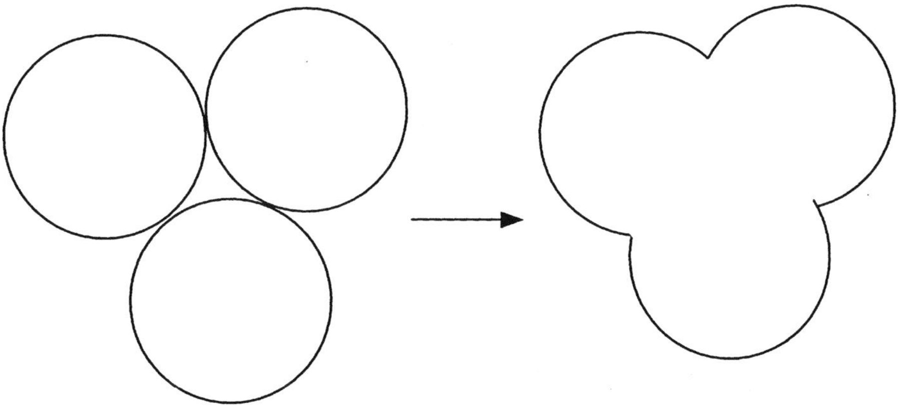

FIGURE 12–*Coalescing of dispersion particles.*

Crosslinking

As water is evaporating from the wet polish, free ammonia is also being lost to the air. Free ammonia is in the formulation, in excess of the amount needed to neutralize the different acids present, because it the formulator must ensure that all of the acids remain completely neutralized as ammonium salts. The ammonium salt of the acids assures that the carboxylates are fully hydrated with water so that the effective MFT of the polymer remains low. This reduced MFT, in turn, ensures film formation with the minimum amount of solvents in the formulation.

Once the free ammonia has evaporated from the formulation it is replaced by ammonia from equilibria with ammonium carboxylate salts. The carboxylate functionality, such as present in the ASR, reverts to being un-neutralized acids, causing the ASR to lose its solubility and precipitate onto the surface of the dispersion particles.

The amine ligands of the latent *tetra*-amino-Zinc complex are also in equilibrium with the aqueous phase (the complex is thermodynamically stabile, but kinetically very labile). As the amine content of the water phase drops, amine leaves the metal complex, (this is shown as the bottom equilibrium of Figure 4) freeing the vacated ligand sites of the metal ion to interact with electron pairs from carboxylate oxygen atoms on the polymer. (These polymer carboxylates do not revert to being free acids because their counterion is a Zinc ion, rather than ammonium ions.)

Dried Film

Once crosslinking has occurred, the final stage of film formation is the continued evaporation of water after it diffuses through the plastic film, until water is no longer a separate phase (the film becomes clear). The polymer dispersion particles have coalesced so there are no longer any discrete particle surfaces. The loss of the particle/water interface means the coalescent is no longer held in position by its affinity for water. Similarly, without a separate water phase, the hydrophobicity of the plasticizer no longer keeps it isolated in the core of what was a dispersion particle. Both types of solvent are now free to diffuse throughout the plastic mass of the nascent or 'green' polish film. At the film/air interface, the coalescent can evaporate (the rate of coalescent loss is augmented by the very large surface area of the film) and a coalescent concentration gradient is established through the depth of the film.

Plasticizers diffuse throughout the film in the same way, and similarly lose their solvating and softening power as the local concentration diminishes. Though plasticizer diffusion also causes migration of these solvents to the film/air interface, their evaporation is much slower than for coalescent because of the plasticizer's affinity for the polymer.

The plasticizing or softening efficiency of solvents is proportional to their concentration. The simple diffusion of the solvents from high concentration regions to an even distribution throughout the film causes the film to harden. The evaporative loss of coalescent further accelerates film hardening.

The dried polish film is now ready to take on the ravages of pedestrian traffic.

Brian T. Cartwright[1]

The Interaction and Performance of Commercial and Experimental Fluorosurfactants and Commercial Floor Polish

Reference: Cartwright, B. T., "**The Interaction and Performance of Commercial and Experimental Fluorosurfactants and Commercial Floor Polish,**" *Technology of Floor Maintenance and Current Trends, ASTM STP 1448*, W. J. Schalitz, Ed., ASTM International, West Conshohocken, PA, 2004.

Abstract: Recently, the leading fluorosurfactant used in floor polish applications has been removed from the market. In an effort to qualify replacements, a thorough study of the use of fluorosurfactants in floor polish applications has been conducted. Water based films have always been subject to temperature and humidity differences in real world applications. Floor polishes are formulated to be useful under these conditions of high or low humidity and high or low temperatures. Certain conditions will also affect the flow and leveling performance of these films. Stress conditions, those on the boundaries of real world examples, can define the performance of a floor polish in today's market. Under these stress conditions fluorosurfactants perform differently and affect the performance and characteristics of a floor polish, including failure.

Keywords: floor polish, floor finish, fluorosurfactant, surface tension

Introduction

Floor polish is a material that is used to protect, enhance, improve cleaning, and improve the safety of vinyl floors throughout the world. It is a material that must be easily applied by a group of individuals with a wide range of skill levels and under a wide range of environmental conditions. It is a material that must both improve the appearance of a floor and make a floor easier to clean. It is a material that is durable to millions of footsteps with only a few very thin coats. Floor polish also provides a uniform, slip resistant surface that keeps those millions of footsteps safe. Floor polish is used in almost every type of business and every type of building. These areas include office buildings, schools, hospitals, shopping centers, grocery stores, and our homes.

Although there are several alternatives on the market today, the majority of people apply floor polish using a string mop. This method is easy to use, easy to teach, and does not require expensive equipment. The use of a string mop does have some variables associated with its use. The amount of floor finish applied to the floor is determined by both the composition of the mop (i.e. the type of fibers used to make a

[1] Technical Manager – Floor Care, OMNOVA Solutions, Inc., 1 Gencorp Dr., Chester, SC 29706-2637.

mop and how those fibers are woven together) and the amount of floor polish that is wrung out of the mop before using.

Different fibers used to make mops have different absorption capabilities and can cause differences in applying floor polish. In addition, other differences in the construction of mops can introduce variables that affect the way a floor polish is applied to the floor. Floor polish formulations must be designed to be used with many types of mop construction.

Substrates affect the application of floor polish. Most commercial floor polishes are designed to be applied over a wide variety of substrates. There are specific examples where floor polish manufacturers have designed a floor polish for a particular type of flooring, but this generally remains a niche market. Some examples of different types of flooring materials where polish is used are vinyl composition tiles (VCT), rubber, terrazzo, stone, concrete and many other types of floors including cork. This presents a challenge to the floor polish manufacturer to produce a floor polish that can be used on many different types of floors. The differences in floors are the porosity, surface tension, texture, composition and resiliency. Even the age of the flooring material can affect the physical properties of the material. Floor polish must overcome all of these differences and apply in a uniform, level coat.

Modern floor polishes are a combination of many ingredients. The largest ingredient is water. Water is an economical and safe way to suspend the components of a floor polish so they can be applied easily. Emulsion polymers are the major component used to form the film that is called floor polish. Coalescing aids and plasticizers help to form a strong, durable film. Polyethylene and/or polypropylene waxes are used to modify the wear and repair properties of the floor polish. Preservatives help to ensure the integrity of the product is maintained over long periods of time. Defoamers are used to reduce the foam in the floor polish and prevent defects in the film caused by excessive foam. Finally, fluorosurfactants are used to affect the flow, leveling, and wetting of a floor polish. All of these materials work together to produce a long lasting, easy to apply, durable product.

For many years, fluorosurfactants have been used in the formulation of floor polishes. Specifically, fluorosurfactants have been used to lower the surface tension of the floor polish. By lowering the surface tension, floor polishes will wet-out low surface tension substrates. If the surface tension of the polish were too high, then the polish would bead up on the surface similar to water beading on a newly waxed car. If the polish "wets-out" then the surface is evenly covered with the floor polish. Fluorosurfactants also affect the flow and leveling of a floor polish. The mechanics of how a floor polish flows over a surface are beyond the scope of this paper. Leveling is simply how level the surface is after the floor polish has been applied to the floor. Immediately after a floor polish is applied with a string mop, an uneven surface of ridges and valleys can be observed in the liquid floor polish. As the polish levels, this texture can not be seen.

There are several application problems that are associated with the fluorosurfactant performance in a floor polish. With no fluorosurfactant, a floor polish will usually not wet-out the surface. With some fluorosurfactants, too much will create excess foaming that can cause surface defects during the application of the finish. Floor polishes that are too viscous can dry before the floor polish has time to level leaving an uneven surface. This same effect can be seen with very thin coats of some floor polishes. Once again, the film dries before it has time to level. Contamination of the floor can also cause floor polishes not to wet or level correctly.

Experimental Section

For purposes of this paper, many of the materials used are not described completely. This is due to the proprietary nature of both the materials and the mixtures discussed.

Floor polishes are mixed by adding water, glycol ether, plastisizer, and fluorosurfactant, and mixing for 30 minutes. Fluorosurfactant was added at a level of 150ppm on an active basis. To this mixture, a commercially available latex polymer is added and mixed for 30 minutes. To the resulting mixture, polyethylene wax emulsion, polypropylene wax emulsion, defoamer emulsion, preservative are added and mixed for 15 minutes. The floor polishes were drawn down the next day.

Drawdowns were completed by using stripped official vinyl composition tile. An amount of 1.70 mL of polish was applied to ½ of an official vinyl composition tile using ½ of a 2"x2" sterile gauze pad, and using a fisher pipetman with a 3.5 mL tip. A control polish of similar composition with the exception of a difference of fluorosurfactant was applied to the other ½ tile. A visual comparison was made to observe surface defects in the polish. At 20% relative humidity, the polishes were recoated every 15 minutes with a total of 4 coats being applied. At 50% relative humidity, the polishes were recoated every 35 minutes with a total of 4 coats being applied. At 80% relative humidity, the polishes were recoated every 75 minutes with a total of four coats being applied. Each coat was visually inspected for surface defects. All four coats were applied during the same day. Polishes with no visual defects were tested with a battery of performance test to ensure the integrity of the polish compared to the control.

Results

Surface tension reduction in floor polish is not solely due to the addition of fluorosurfactant. The other raw materials that are used to manufacture floor polish contribute significantly to the surface tension reduction. See Table #1 Surface tension reduction in floor polish. Another significant contributor to surface tension reduction is the addition of the plasticizer Tributoxy Ethyl Phosphate sold commercially as KP-140 or TBEP. The addition of emulsion polymers and wax emulsions reduce surface tension primarily due to a slight excess of hydrocarbon surfactant(s) used to emulsify the respective polymers and waxes.

As stated in (Table #1), the combination of Water, Glycol Ether (Coalescing aid), Plasticizer, and Latex Polymer has a low surface tension measurement that is similar to the surface tension measurement that includes fluorosurfactant. Even with all of the necessary components except fluorosurfactant the surface tension measurement should be low enough for good performance, but these polishes have uncertain performance.

Table 1—*Surface tension reductions in floor polish.*

Floor Polish Component	(g)	(g)	(g)	(g)	(g)	(g)
Water	100	93.40	60.03	53.6	48.26	48.26
Glycol Ether		6.60		3.41	3.41	3.41
Plastisizer				2.99	2.99	2.99
Latex Polymer			39.97	39.97	39.97	39.97
Wax Emulsion					4.75	4.75
Defoamer					.02	.02
Fluorosurfactant (30% active)						.05
Surface Tension (mN/m)	75.6	72.7	43.4	29.6	30.3	27.9

Several commercially available fluorosurfactants and several experimental fluorosurfactants were tested to determine the surface tension of the floor polishes using different fluorosurfactants. Two basic fluorosurfactant chemistries were employed. The fluorosurfactants used were based on substituted telomer alcohol chemistries with either an ionic or nonionic hydrophile and short chain fluorinated poly(oxetane) chemistry. The latter has a hydrophobe between two hydrophiles. Both anionic and non-ionic surfactants were used. Typical surfactants are thought of as having both a hydrophobic segment and a hydrophilic section. This geometry allows the hydrophilic section to associate with the aqueous portion of the mixture and the hydrophobe to associate with hydrophobic portions of the floor polish formulation or at the air/water interface. The latter disturbs the surface causing changes in the surface tension of the mixture. There are many surface-active agents in a floor polish formulation. The fluorosurfactant must not exclusively compete with these other agents on the surface; it must be complementary. Both chemistries used in these experiments have a hydrophile to aid in the solubility of the materials in aqueous systems, and a fluorinated portion that will partition itself at the air/water interface to affect the surface tension (wetting), flow, and leveling properties.

Once the surface tension measurement was obtained, the floor polishes were drawn down to evaluate performance. Four coats of finish were drawn down to determine the ability of the floor polish to wet and level during experimental conditions on official vinyl composition tile and over subsequent coats of the polish itself.

There are floor polishes that have low surface tension measurements and poor performance. All of the samples described in Figure #2 were made with the same formulation with a single exception of a difference in fluorosurfactant. Since all of the other materials were held constant there are several conclusions to make. There can be chemical interaction that prohibits the material from working in an appropriate manner. This chemical interaction could be electrostatic, steric, or miscibility with the components of the floor polish. To avoid electrostatic interactions, only non-ionic or anionic fluorosurfactants were used in this experiment. Steric interactions could be caused by the way the materials compete on the air, polish interface.

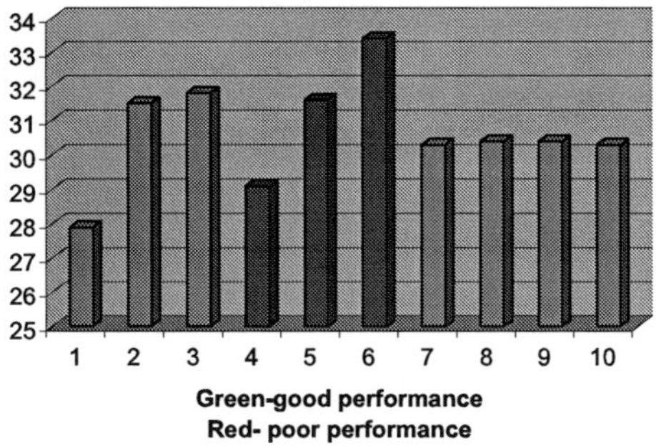

FIG. 1—*Surface tension measurement of floor polishes with different fluorosurfactants and their performance.*

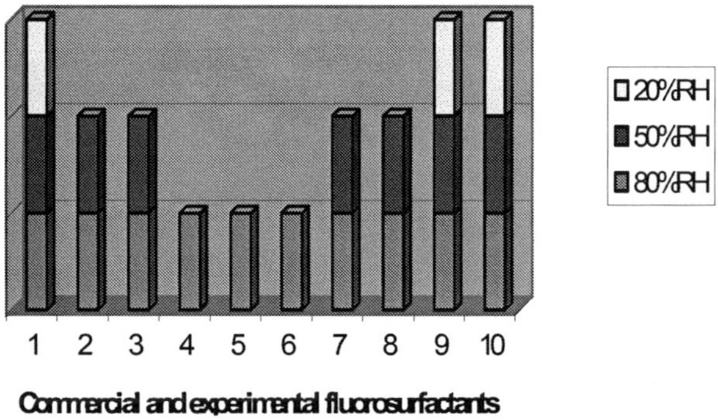

FIG. 2—*Fluorosurfactant performance under different relative humidity environments.*

These inconsistencies with conventional wisdom prompted a different approach. As stated earlier, leveling problems with floor finish can sometimes be attributed to finishes drying before they have time to level out completely eliminating defects. Experiments were performed by drawing down floor polishes at different relative humidity with different fluorosurfactants. At high relative humidity (80%RH), several polishes had acceptable leveling without fluorosurfactant. All of the fluorosurfactants tested had acceptable performance at 80% relative humidity. At 50% relative humidity, seven polishes had acceptable performance. At low relative humidity (20%RH), only three polishes had acceptable performance (this was not exclusive to a single type of fluorinated chemistry). Only polishes that leveled well at high and normal humidity performed well at low humidity. None of the polishes/fluorosurfactants tested performed at low humidity and not at high humidity.

The difference in relative humidity largely affects the dry time of the floor polish. Floor polishes dry in less than 15 minutes at 20% relative humidity. Floor polishes dry in approximately 35 minutes at 50% relative humidity. And at high humidity (i.e. 80%RH), floor polishes take 45 minutes to over an hour to completely dry. At higher relative humidity, floor polishes have longer to level than floor polishes applied at low relative humidity. Therefore, there is a time component to the effectiveness of fluorosurfactants in aqueous floor polishes. In addition, it is highly recommended that the development process of floor polishes include performance at low humidity conditions to evaluate the effectiveness of the fluorosurfactants.

Conclusions

The interactions of fluorosurfactants in floor polish are very important. Without proper complimentary or synergistic interaction, floor polish would not perform to the necessary level of performance demanded by the industry. There are interactions that go beyond simple surface tension measurements to determine the success of a fluorosurfactant in a given floor polish. There are low surface tension floor polishes that do not perform adequately. In addition, there are other interactions and properties that can help to predict success with a floor polish. Time is a critical element in the mechanism of fluorosurfactants and their performance. Only effective fluorosurfactants can help to level a floor finish in the shortest time available. Floor finish must be able to perform at all practical levels of relative humidity since they are routinely applied at various conditions.

Acknowledgement

The author thanks Mrs. Joyce Wages, Mrs. Sandy Williford, Mr. Russ Craig, Drs. Bob Medsker, Rick Thomas, and Ted Del Donno for many helpful discussions. The author acknowledges gratefully suggestions made by the reviewers.

Theodore Tysak[1]

Water-Based Acrylic Concrete Seals

Reference: Tysak, T., **"Water-Based Acrylic Concrete Seals,"** *Technology of Floor Maintenance and Current Trends, ASTM STP 1448,* W. J. Schalitz, Ed., ASTM International, West Conshohocken, PA, 2004.

Abstract: Stringent environmental regulations have forced the concrete finishing market to reduce volatile organic compounds (VOCs). Formulators use water based polymers to reduce these emissions. In this paper, the chemical composition of commercial aqueous acrylic latex concrete seals, and their performance is presented. A discussion of efflorescence and its prevention, as well as concrete sealer application and maintenance procedures are also presented.

Keywords: concrete, efflorescence, film formation, acrylic, chemical resistance

Introduction

Structural concrete functions as flooring in many industrial and commercial buildings such as warehouses, fire houses, factories, retail home centers and parking garages. Concrete serves well in this role because it has the requisite compressive strength and durability. Using structural concrete for flooring surfaces is practical because it is cost-effective; however, uncoated concrete presents two major shortcomings as a flooring material. First, it is porous, and second, it is relatively brittle with an uneven surface pattern.

Permeability is a disadvantage because it makes concrete highly susceptible to deep staining. Powdery solids such as fine soil, and soot and liquids such as gasoline, grease, oil and water spilled on concrete tend to work their way deep into its surface. The embedded materials can appear as unsightly discolorations and are difficult to remove. Some of the staining materials, such as grease, oils and surfactants, can also make the floor slippery when they become lodged in the concrete surface.

The fragility and irregular surface pattern of concrete lead to a different problem—development of dust. Heavy traffic passing over the concrete surface causes

[1] Principal Scientist, Floor Care, Rohm and Haas Company, 727 Norristown Road, Spring House, PA, 19477.

tiny surface imperfections to flake-off and be released as dust particles. Strictly a surface phenomenon, the dust generation does not affect the integrity of the floor structure, but it unquestionably contributes to the airborne particulates in the atmosphere of the work area. Standards enacted by the regulatory agencies set limits for the amount of dust particulates that can be present.

To protect concrete flooring from staining and dusting, industrial maintenance crews customarily employ sealers. Part of the applied sealer flows into the surface pores of the concrete; the remainder forms a thin layer atop the surface. Once in place, the sealer cures or dries into a hard, tough protective film that prevents wear of the concrete and shields its surface from the abusive forces of mechanical and pedestrian traffic that crosses over the concrete floor. Additionally, the sealer helps in the prevention of efflorescence, a common problem of concrete surfaces.

The focus of this paper will be on the use of aqueous-based acrylic latex technologies as film formers for concrete coating applications with an emphasis on two areas: efflorescence prevention and final coating properties.

Film Formation of Acrylic Latex Sealers

An acrylic latex consists of discrete polymer particles dispersed in water; the individual latex particles are themselves made up of very long polymer molecules that are highly entangled. A brief review of the film formation mechanism for acrylic latex sealers is appropriate since this information will aid in understanding how acrylic latex sealers are used to coat concrete preventing efflorescence and influence final coating properties.

Solvent based acrylic polymers and other solvent borne chemistries form a film by the simple evaporation of the solvent. Film formation from an aqueous acrylic latex is considerably more complex, and takes place in three discrete steps. In Stage I, water evaporates from the latex surface, gradually forcing the spherical latex polymer particles into a closely packed array. In Stage II, initial coalescence occurs as interstitial water evaporates and the polymer particles undergo deformation due to interfacial pressures and surface tension affects. Water loss by free surface evaporation is quite rapid, and the solids increases from an initial 40 % to 85% or so at Stage II. Thereafter, the loss of remaining water and coalescing solvents occurs by diffusion, a much slower process. Ultimately, this pressure leads to a transition stage in which the particles assume a closest packing hexagonal array. Finally, Stage III is reached in which the interparticle boundaries lose some definition, and the polymer chains within each particle intermix to a certain extent to form a smooth, homogeneous film. Figure 1 depicts this process as an idealized model and Figure 2 shows an electron photomicrograph of an actual acrylic latex film.

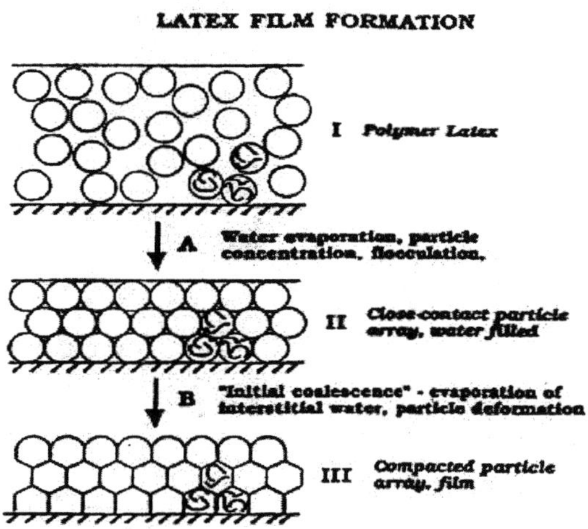

Figure 1 - *"Idealized" representation of acrylic latex film formation.*

When applied, the formulated acrylic sealer will penetrate the porous concrete surface to afford good adhesion and form a continuous film. Another function which the sealer serves is helping prevent efflorescence.

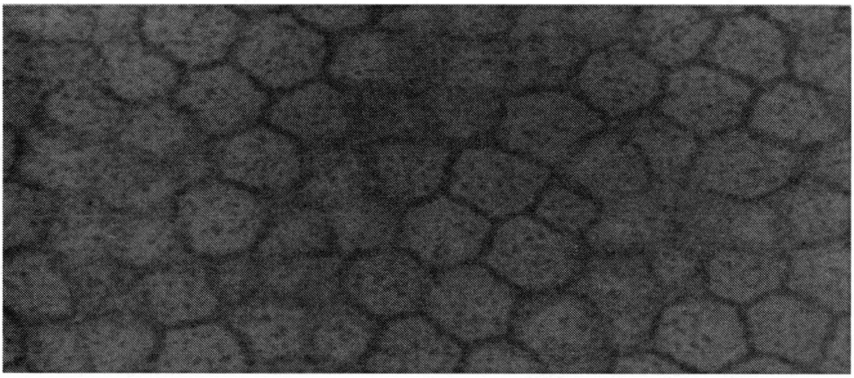

Figure 2 - *Transmission electron micrograph of an acrylic latex film (typical concrete sealer composition, formulated) "stained" showing the bulk film structure. Average particle diameter was determined to be approximately 100 nanometers.*

What is Efflorescence?

Efflorescence is a naturally occurring phenomenon that results from the hydration of cement, a component of concrete. Efflorescence is the deposit of salts on the surface of concrete, which is caused by leaching of minerals from the substrate by water. Thus, it is seen to occur on everything from highways and bridge abutments to such manufactured concrete articles as pavers and concrete roof tiles. While the deposits of white salts resulting from efflorescence on the surface of naturally colored concrete are not particularly objectionable, their presence on colored (pigmented) concrete substrates is commonly considered objectionable from an esthetic standpoint. Preventing the efflorescence from occurring is an important consideration.

Primary and Secondary Efflorescence

Primary efflorescence occurs typically over the first several hours after the concrete is mixed and poured into place. As cement hydrates, significant amounts of salts - principally calcium hydroxide - are released as aqueous solutions and transported through capillaries to the surface of the concrete.

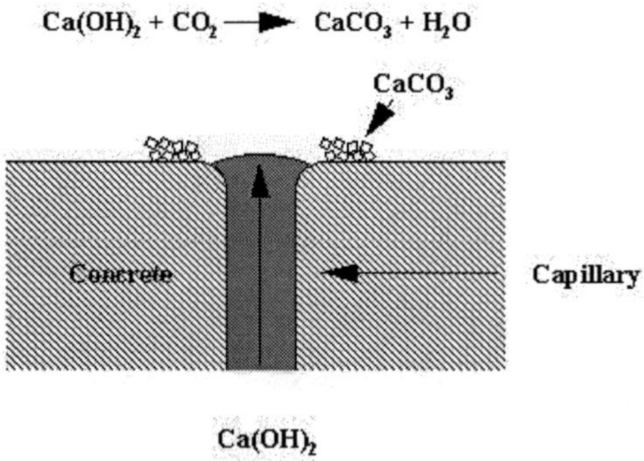

Figure 3 - *Representation of formation of primary efflorescence in presence of excess water.*

As depicted in Figure 3, the calcium hydroxide then reacts rapidly with atmospheric carbon dioxide to form an insoluble white deposit of calcium carbonate.

$$Ca\,(OH)_2 + CO_2 \rightarrow H_2O + CaCO_3 \text{ (insoluble)}$$

As the concrete substrate cures, less water is present at the surface and some of the white calcium carbonate resulting from primary efflorescence remains at or just below the concrete substrate surface within the capillaries of the structure instead of appearing as a uniform whitish layer of crystals over the entire surface of the substrate (Figure 4).

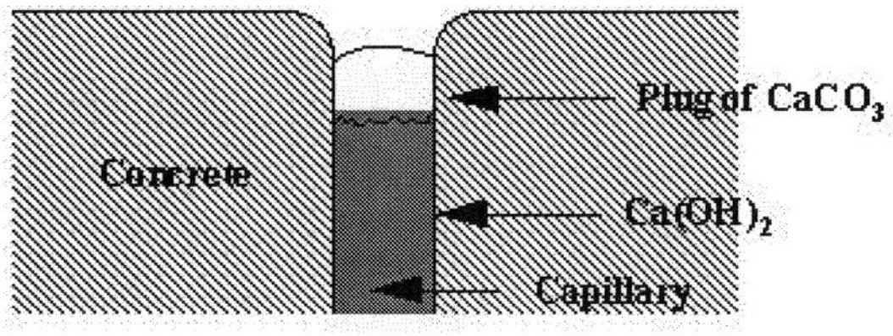

Figure 4 - *Formation of primary efflorescence in concrete (near the surface) with minimum water.*

Secondary efflorescence results when water penetrates the capillary structure on the concrete substrate surface and extracts soluble salts from the body of the concrete (Figure 5). These salts may consist of calcium hydroxide arising from the continuing process of cement hydration, which will result in a surface deposit of insoluble calcium carbonate. Eventually, these deposits will ultimately be converted to soluble calcium bicarbonate upon reaction with additional atmospheric carbon dioxide and water as illustrated by the following chemical equation:

$$CaCO_3 + CO_2 + H_2O \rightarrow Ca\,(HCO_3)_2 \text{ (soluble)}$$

In addition, water-soluble salts such as the carbonates and sulfates of sodium, potassium, and magnesium, along with chlorides of calcium and other metals, may also be deposited on the surface. If the concrete substrate is exposed to water (rainfall on exterior concrete), these salts will eventually be washed away fairly rapidly until the supply of them in the concrete substrate accessible to water is exhausted. However, if the substrate is exposed to limited amounts of water, or no water at all, such as is the case with interior floors, the process of salts leaching from the concrete substrate and depositing on the surface may occur over and over again for a lengthy period of time.

Unlike primary efflorescence, secondary efflorescence may occur for weeks - indeed months - following the initial installation of the concrete substrate. And unlike primary efflorescence, which tends to impart a uniform whitening to the surface, secondary efflorescence may occur sporadically and randomly over this extended time frame, resulting in a very objectionable, mottled appearance in the concrete substrate.

The Use of Acrylic Sealers to Prevent Efflorescence

Over the years, a number of approaches have been tried to reduce or eliminate efflorescence, including special cements, certain aggregates, fatty acid soaps such as stearates, and silicone additives. None of these have proven entirely satisfactory; acrylic latex sealers can be used as coatings to prevent efflorescence.

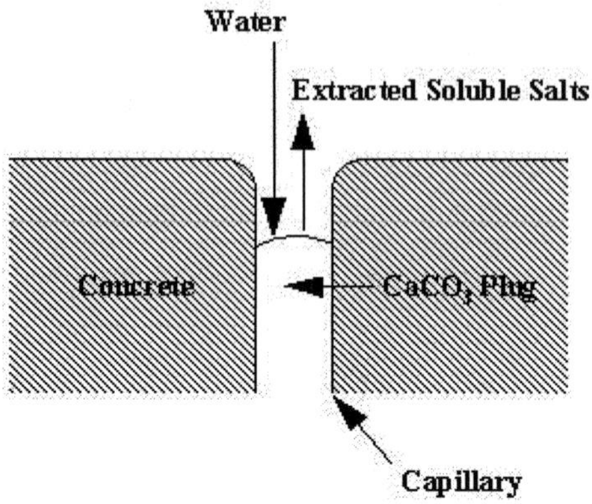

Figure 5 - *Schematic representation of secondary efflorescence.*

As exterior coatings for cementitious and concrete substrates, both solvent-based and water-based acrylic polymers are widely used because acrylic polymers are transparent to ultraviolet light, and hence are very durable to exterior weathering. Besides greatly reduced VOC and flammability hazard relative to solvent-borne sealers, water-based acrylics confer other advantages. Aqueous acrylic latexes have much higher molecular weights that make tougher and more elastic films, which withstand the stresses of repeated temperature cycling.

How Acrylic Sealers Prevent Efflorescence

Typically, concrete requires 27 – 30 days to fully cure. Properly applied, a formulated acrylic latex polymer will prevent efflorescence by forming a thin, continuous film over the surface of the concrete substrate. It can act to reduce primary efflorescence by blocking the escape of calcium hydroxide to the surface of the concrete during cure, while allowing carbon dioxide to pass through the film to form a plug of calcium carbonate within the numerous capillaries present in the concrete (Figure 6). When an acrylic latex sealer is applied early (usually within 24 hours after casting or pouring the concrete), it prevents primary efflorescence by not allowing liquid water to enter the concrete substrate but will allow water vapor to leave the concrete surface. An acrylic sealer applied in this fashion will further benefit the concrete substrate by allowing the concrete surface to develop better surface hardness characteristics because the cement hydration process is slowed down. Additionally, the film acts to prevent secondary efflorescence by forming a barrierwhich prohibits various forms of moisture from penetrating the capillary network at the concrete surface and extracting soluble salts. Refer to the schematic in Figure 7.

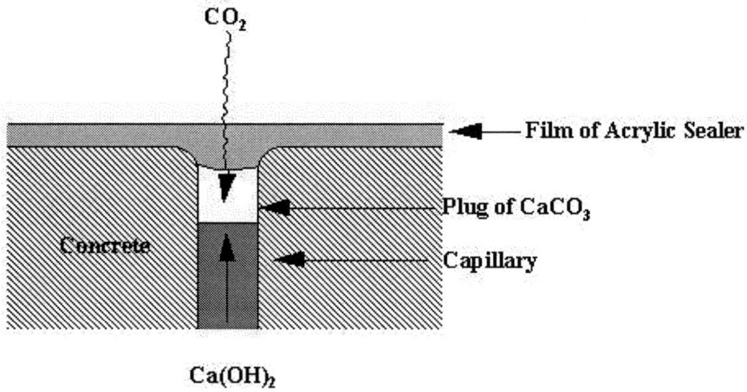

Figure 6 - *Schematic of how acrylic latex sealer reduces primary efflorescence by blocking migration of calcium hydroxide [Ca (OH₂)] to the concrete surface while allowing carbon dioxide [CO₂] to pass through to form plug of calcium carbonate [CaCO₃] in the capillaries.*

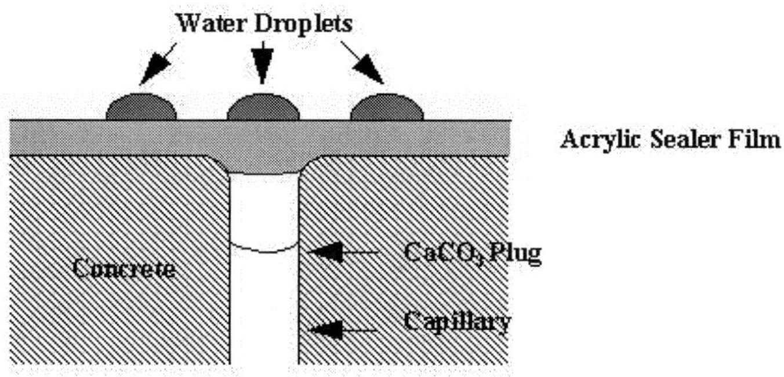

Figure 7 - *Schematic of acrylic latex sealer film preventing secondary efflorescence by acting as a barrier to water.*

Final Sealer Properties

An acrylic latex sealer needs a variety of other properties to function satisfactorily. First, it must have a rheological character that permits it to flow into the concrete pores but not so completely that the sealer is totally absorbed. Pore sizes in concrete substrates vary tremendously and can be as small as 20 to 50 nanometers. To protect the concrete a good portion of the sealer should remain on the surface (Figure 8).

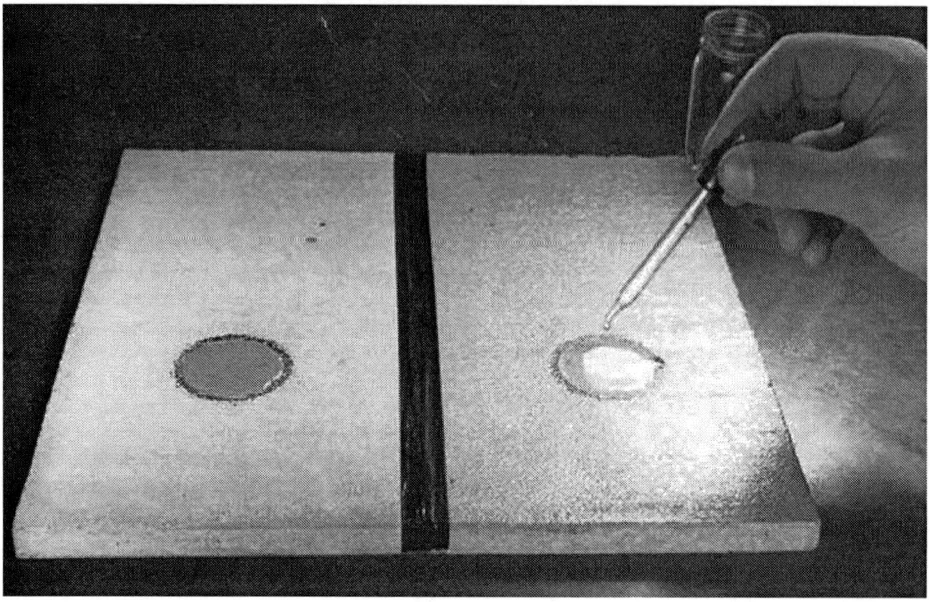

Figure 8 - *In a laboratory test, a concrete slab illustrates the protective function of a one-pot sealer based on an acrylic latex vehicle. The portion on the left has not been finished; the portion on the right has been sealed. A small amount of liquid has been placed on each surface. The unsealed concrete readily absorbs the liquid, but the sealer on the other side of the slab prevents the liquid from penetrating the surface.*

Second, the sealer needs chemical and water resistance to endure the erosive effects of spilled materials and to withstand the highly caustic concrete surface. Third, it should be hard and durable enough to hold up when exposed to conventional mechanized and pedestrian traffic. Fourth, it should have gloss for a desirable appearance (Figure 9).

The industry has employed many different materials as sealers but all have certain shortcomings. For example, formulations based on two-package solvent-borne curing systems such as epoxies and some urethanes provide excellent durability and resistance properties. However, they are very expensive, their curing mechanisms are sensitive to relatively minor fluctuations in ambient temperature and humidity, and their combustible and hazardous solvent carriers make it difficult to meet increasingly restrictive government regulations. In addition, as two-package systems, they require more careful preparation and handling than one-component products. This extra handling infers higher costs in terms of labor.

Figure 9 - *The left hand portion of this concrete slab has been sealed with an aqueous-based acrylic polymer designed for resilient floor sealer applications; the right-hand portion with a sealer based on an acrylic latex specifically developed for concrete coating applications providing a clearly glossier surface.*

Products based on aqueous urethanes and solvent-borne acrylics have also been formulated into sealers with varying degrees of success. In general, these polymers offer good resistance properties and durability at less cost than the solvent-borne urethane and epoxy products. Unfortunately, these vehicles tend to be very expensive. In addition, solvent borne acrylics face the same kinds of regulatory limitations confronting all solvent systems.

The desire for more cost-effective and environmentally acceptable sealers has attracted manufacturers and end-users to fairly inexpensive one-package aqueous acrylic systems. These products supply acceptable durability and resistance properties in many other allied coatings applications.

Thermoplastic acrylic emulsion polymers containing the right balance of monomers can be formulated into high quality concrete sealers. Typical commercial acrylic latex polymers are usually copolymers of several acrylate and methacrylate monomers, the combination of which offers a wide range of strength, elasticity and other key properties (Figure 10). The presence of certain types of functionality in the polymer along with the glass transition (Tg) and molecular weight of the polymer backbone imparts specific characteristics to the films of concrete coatings such as chemical resistance and adhesion; typically acrylates and methacrylates that contain 'nitrile', 'sulfo', and 'phospho' moieties are employed. The performance of these thermoplastic emulsion polymers match or exceed the performance of aqueous urethanes, and aqueous/urethane combinations – and at much lower cost. Furthermore, the aqueous nature of these acrylic latex polymers allows the user to avoid the health and safety issues normally present with solvent-borne systems, particularly when applied in areas with limited ventilation.

As far as performance is concerned, one component sealers based on aqueous thermoplastic emulsions display a good balance of properties. These formulations have good leveling and wetting, resistance to water and a wide variety of chemical agents, and good film toughness. Proper flow and leveling ensures that enough sealer will penetrate the pores of the flooring to produce satisfactory adhesion. Very good resistance properties enable sealers based on the thermoplastic acrylic emulsions to withstand the destructive assaults of common industrial chemicals such as acids (Figure 11), alkalis, oil, detergents, and gasoline (Figure 12).

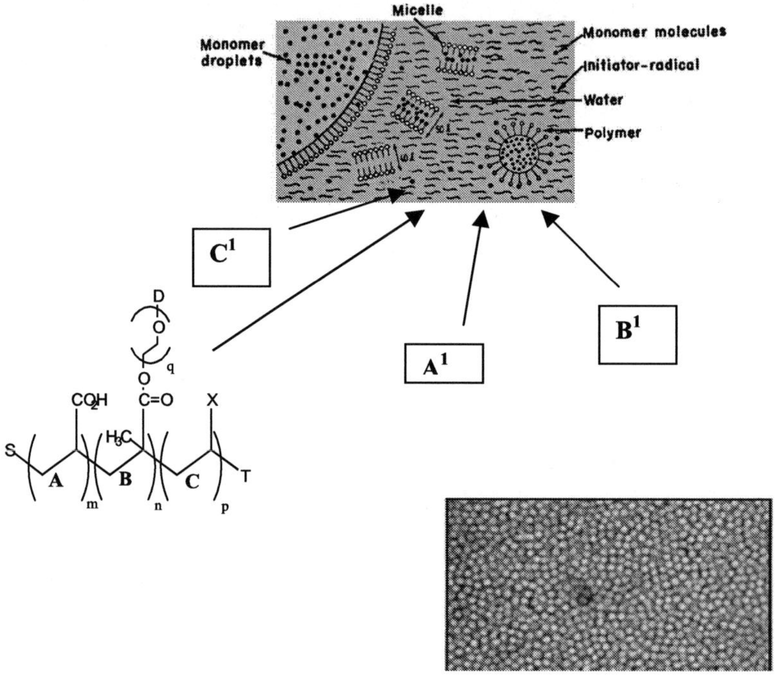

Figure 10 - *Schematic representation of acrylic latex technology. Water serves as the main carrier for the latex system. Surfactants are present in the water, which form micelles. The monomer ingredients upon being charged to the water enter the micelles where the polymerization reaction takes place via a free-radical process. The range of compositions is abundant. Monomer type as depicted by the 'A', 'B', 'C' designations and amount (designated by 'm', 'n', 'p') need to be chosen accordingly from the available acrylate/methacrylate monomers available to the manufacturer. Other unique functional groups (depicted as 'X' and 'D') are typically incorporated to provide some of the key properties required of a concrete sealer: chemical resistance and adhesion. The bottom picture is a photomicrograph of actual acrylic latex particles typically used in the formulation of concrete sealers. The photomicrograph was generated by taking a droplet of the aqueous latex and drying it on a glass slide. A high-powered optical microscope was used to obtain the picture.*

Additionally, the one-component aqueous-based sealers have the integrity to hold up to heavy industrial traffic for reasonably long periods of time.

The aqueous acrylics generally have acceptable resistance and durability properties because of the chemical nature of the acrylic backbone. An acrylic backbone is comprised entirely of C-C single bonds (Figure 10) that are relatively inert and not susceptible to hydrolysis or chain scission as with other linkages, for example, those based on amides or ethers. It is important to note that the acrylic backbone composition must be designed appropriately to meet the requirements of a coating for concrete. Acrylic technology is prevalent in the marketplace, and not all compositions will meet the unique requirements for sealing and protecting concrete surfaces. Figure 11 and Figure 12 show the results of an acrylic technology (a traditional resilient floor tile sealer based on an acrylic latex) that was not designed for concrete coating applications.

Figure 11 - *The type of sealer used for specific applications is critical. In this instance, acid was applied to a concrete slab, half of which was sealed with a sealer coating not designed for concrete and half with a high performance, one component, aqueous-based concrete sealer. The acid has eaten through the coating on the left side, but the side protected with the high performance aqueous based sealer remains intact.*

As indicated previously, aqueous acrylic polymers used as primary vehicles for water based seals are inherently nonflammable and basically non-hazardous. Consequently, they are much easier and safer to formulate, handle, and apply than products containing solvent-borne polymers. The economic benefits of thermoplastic acrylic emulsions stem from their lower cost and the recoatability of sealers into which they are formulated. These emulsions are low cost because the acrylic polymer is reasonably inexpensive and uses water as a carrier. In contrast, both the polymer and solvent carrier of curing systems and some of the solvent-borne systems are expensive.

Other systems such as aqueous urethanes also rely on water as an inexpensive carrier, but the polymer or vehicle component is very costly. The reverse is true for the solvent-borne acrylic polymers.

Recoatability offers economic advantages in a different area. Cured coating systems that have worn to a point that protection of the substrate has been compromised must be repaired or replaced. Coating replacement is a tedious, time-consuming procedure that requires complete removal of the existing sealer—a lengthy and difficult task (particularly if the sealers are cured reactive materials or if they have penetrated the concrete substrate deeply), conditioning the floor, rinsing, and, finally, application of a new coat.

Figure 12- *Gasoline proves to be a strenuous test for sealers. This concrete slab has been sealed on the left with a sealer typically used for vinyl composition tiles and on the right with a high performance, one component, aqueous sealer designed for concrete. Gasoline has been applied to the bottom half of each side. The "vinyl composition" tile sealer has been completely dissolved by the gasoline but the high performance, one component, aqueous-based sealer has not been affected.*

The time required for the entire process may run anywhere from three to seven days, an interruption that can be most inconvenient for the operator of the facility. In comparison, repair involves simply cleaning the worn areas and applying new coat sealer. A maintenance crew can perform this work in a short time span, typically within several hours, with obvious savings in labor and less inconvenience for the facility owner.

Repair is possible, however, only when a sealer is recoatable. For a sealer formulation to have this property, a new coat must adhere well to the old layer lying on the floor or existing in the concrete pores. Most sealers employing curing systems such as epoxies and urethanes lack the inter-coat adhesion necessary for recoatability. Consequently, they must be totally replaced when they become worn or damaged. In comparison, one component, aqueous acrylic based concrete sealers typically provide inter-coat adhesion, faster dry times and consequently recoat extremely well.

Formulating and Application

The aqueous, one package nature of thermoplastic aqueous-based vehicles makes formulation of sealers easy and straightforward. Formulation and processing techniques are similar to those encountered in the resilient floor finish industry; simple addition and agitation techniques are all that is required. End user requirements will dictate the characteristics required of the sealer and influence other raw material component choice. Care must be exercised with other materials selection because they will influence flow, leveling, wetting, gloss, color, and wear properties of the sealer.

Achieving good film formation is the most important aspect of a concrete sealer's commercial success. Poor film formation can lead to many problems. Some can go unnoticed for months (stress cracks from temperature changes); others, like poor chemical and water resistance, become apparent in a very short period of time. Film formation is greatly dependent on the evaporation rate of water from the film (Figure 1) and the retention of solvents within the film. Choosing the proper amounts and types of solvents, such as coalescents and plasticizers to aid film formation and influence the minimum film formation temperature, is probably the most difficult aspect of formulating a concrete sealer using an aqueous acrylic latex.

Solvents can be grouped into three different categories depending upon where they partition ("live") in a given latex system. The partitioning of coalescents and plasticizers depends on many things, such as the solvents miscibility with water, the acrylic latex composition, molecular weight and morphology. In general, a latex particle can be separated into three distinct regions (Figure 10): the "A^1" region or interior of the latex, the "B^1" region or exterior of the latex particle, and the "C^1" region or water phase. "A^1" region solvents, typically aliphatics and aromatics, are very hydrophobic and will typically swell a latex, however, they are inefficient in helping the latex form a film (reducing the minimum film forming temperature of the polymer). "A^1" region solvents, usually referred to as plasticizers, are non-volatile and will contribute to the adhesion of the formulated concrete sealer by allowing the latex particles to deform readily and fill the pores of the concrete. Concrete is a porous substrate extracting solvents from coatings and judicious use of "A^1" solvents are effective in assisting the film formation process. "B^1" region solvents (glycol ethers/alcohols) are generally the most effective coalescents (effective for providing coherent and uniform films) due to their presence in the outer region of the latex particle where the individual particles come in contact with each other (refer to Figure 1). The amount of swelling a "B^1" region coalescent imparts to a latex is associated with its degree of water solubility. "C^1" region solvents (hydrophilic glycols and ether/alcohols) are completely water miscible and less effective coalescents due to their presence in the water phase. They do not partition into the latex and behave more as diluents for the latex formulation. Both "B^1" and "C^1" region coalescents are essential to the formulation for their ability to reduce the surface tension of the concrete sealer coating contributing to better flow and leveling. With most formulations, combinations of "B^1" and "C^1" type solvents are used to balance film formation properties.

Formulating waterborne concrete coatings requires an understanding of the interactions between the latex, solvents and other formulation additives that may be required.

Application Methods

An important feature of one component, aqueous acrylic polymers for use as concrete sealers is that they can be formulated to possess excellent rheology. When used with the appropriate thickener (if desired), acrylic latex formulations have a low shear viscosity and a high shear viscosity. Those characteristics give coatings applied by brush and roller excellent flow, leveling, and film build. As a result, aqueous acrylic-based concrete sealer films formulated to have the appropriate rheology are smooth, uniform, and can be made to be reasonably thick, if this a preference. Another benefit to using aqueous acrylic one component based concrete sealers and appropriate rheology modifiers is if coatings will be applied by airless sprayer; control of foam is much easier in these formulations.

Critical Conditions for Successful Use of Acrylic Latex Sealers

While application is relatively easy and uncomplicated, successful placement of the sealer depends very heavily on the proper preparation of the concrete beforehand. The flooring substrate must be free from oil and oleo-resinous stains or the sealer will not adhere well to it; this is particularly important for pre-existing concrete surfaces. Unsoiled, unsealed concrete with embedded dirt requires thorough scrubbing with specially formulated degreaser/cleaner or detergent solution before the concrete can be sealed. In general, it is essential that a film of adequate thickness be uniformly applied to the substrate surface to function properly. Correct application should yield a dry film thickness of between one – three dry mils. The latex must be formulated with the proper type and level of coalescent to assure good film formation at the lowest temperatures that will be encountered during the application. With properly designed aqueous acrylic latexes, it appears that less preparation of the substrate is required as compared to solvent-borne systems, particularly those based on crosslinking chemistries (polyol –

polyisocyanate, epoxy-amine and moisture cure) . Before these solvent-borne sealer systems can be applied, not only must the concrete flooring surface be scrubbed with a cleaner, but it must also be acid etched and rinsed. Empirical evidence compiled from product use in the marketplace so far suggests that formulations based on water-borne acrylic emulsions tend to work well foregoing the acid etch step. Users of water-borne systems would expend much less time and labor preparing the flooring substrates than they do with the crosslinking solvent systems by not having to acid etch. Acid etching may be required if efflorescence was determined to be present on the concrete surface.

Summary

Aqueous acrylic latexes as vehicles for concrete sealers offer excellent resistance to chemicals, wear, and solvents; good leveling; and good film toughness and aid in the prevention of primary and secondary efflorescence. They provide these qualities at considerably lower cost than the solvent and aqueous urethanes, epoxies, solvent-borne acrylics, and acrylic/urethane combinations currently used by the market. Water-based acrylic polymers offer particular advantages in terms of reduced fire hazards, low odor and ease of cleanup. Their successful use depends on achieving good film formation by proper selection of acrylic latex, and using proper formulation and application techniques.

COATING MAINTENANCE AND STATIC COEFFICIENT OF FRICTION

Stuart Hughes[1]

Coefficient of Friction - An Overview of Floor Surfaces, Polishes and Maintenance Interaction

Reference: Hughes, S., "**Coefficient of Friction, An Overview of Floor Surfaces, Polishes and Maintenance Interaction,**" *Technology of Floor Maintenance and Current Trends, ASTM STP 1448*, W. J. Schalitz, Ed., ASTM International, West Conshohocken, PA, 2004.

Abstract: Floor care and slip resistance has been fundamentally accepted for centuries. The advent of scientific investigation of specific interactions between flooring materials and traction is a relatively new subject, being pioneered by Hunter during the 1920's and continued by James during the first half of the 20th century. This work established the standard utilized by Underwriters Laboratories, Inc., and eventually to the standard by ASTM D 21, the committee on Polishes. Continued scientific study, correlated with actual field experience solidified the importance of this work and established the benchmark for determining the static coefficient of friction of a polished surface, as well as creating a means for examining the relative slip resistance of other surfaces intended for pedestrian passage.

Keywords: slip resistance, polishes, stick-slip, viscoelastic, static coefficient of friction, flooring, James Machine, resilient

Introduction

To understand better the complexity of slip resistance, it is important to review the science involved in the various aspects of the majority of interactions that play a role in creating a walking surface that is firm, stable, and slip resistant. Among the topics to be discussed is an overview of the importance of friction and its measurement through the use of the James Machine, as well as the interaction of flooring surfaces and maintenance coatings.

What is a Safe Floor Finish?

The subject of safety in floor finishes is admittedly one of an extremely controversial nature. Slipping accidents are sometimes difficult to explain and oftentimes there are psychological as well as mechanical elements involved. People have long associated a floor

[1] Technical Service Director, Hillyard Industries, Inc., 302 North 4th Street, St. Joseph, MO, 64501.

floor of glossy appearance with slipperiness, although as we now know, there need be no relationship at all [1].

Those words, written by Sidney James of Underwriters Laboratories in 1944, are as true today as they were when originally published. Contributing to the complexity of investigating the interaction between walking surfaces and creating a reasonably safe walkway are the intricacies involved in the process of walking. As we evolved from the stability of four legs to bipedal ambulation, efficient locomotion assumes an adequate frictional interface where the foot (or shoe sole material) meets the floor.[2]

It is generally accepted from repeated biomechanical studies of walking, that human locomotion is possible only when the foot/floor-walking interface provides a sufficient frictional force. The frictional force available must be adequate enough so that the foot remains stationary, relative to the floor surface when horizontal forces are applied to either push forward or stop the walking body by the leg. If the frictional force is insufficient, the horizontal force will cause the foot to move, relative to the floor, and this is known as a "slip." Because the center of gravity of the walking body is unsupported during a slip, unless it is caught or brought under control, a slip can result in a fall. However, if the center of gravity is successfully brought under control after either a slip or a trip, the incident is called a "stumble."

The frictional force exerted at the walking interface to oppose the horizontal forces of walking is a property of the two surfaces, which meet at the walking interface. It is a function of the chemical compositions and physical structures (particularly roughness and planarity) of both of the surfaces that make the interface. Since a polish-coated floor is very often one of these surfaces, the frictional properties of polish films have been of great importance to the floor polish industry.

There are a number of universally accepted, pertinent facts about friction as it relates to human locomotion. Scientists have studied the force of friction by closely examining those instances where its effect has been reduced to zero. That is, they examine and vary the forces that are acting on an interface at the instant when slipping just begins to occur. In this instant the frictional force keeping the two surfaces of the interface from moving relative to each other is exactly balanced by the tangential force, which is trying to move them.

The frictional force available at the walking interface is directly proportional to the vertical force holding the interface together. This is called the "normal" (or vertical) force and is provided by the mass or weight of the pedestrian. If an increasing horizontal (or tangential) force were applied to the interface, at some point it would become greater than the restraining frictional force and the two surfaces of the walking interface would move relative to each other. At some point, by increasing the horizontal force, it would overcome the frictional force and the foot would accelerate away (slip) from the point of contact. Higher horizontal forces arise when a pedestrian is accelerating, or running, or sharply changing direction of travel. The same slip result would occur if, instead of increasing the horizontal force, we would keep it constant and now decrease the vertical

[2] For walking on level (horizontal surfaces), a gait cycle occurs between the heel strike of one foot, and the subsequent heel strike of the same foot. The process of walking includes the intermediate steps that occur during the gait cycle, – the toe-off, swing phase, heel strike, and stance phase. Temple, J., *The Staircase Studies of Hazards, Falls, and Safer design*, MIT Press, Cambridge, MA, 1992.

force. Because there is now less force holding the interface together, less horizontal force will be required to cause the surfaces to slip. (More on this topic is discussed later).

Is the Floor Too Slippery?

Is the floor too slippery? One way to find out is to wait for people to fall; the better method is to test the coefficient of friction of the floor finish [2]. Those words, written by Stuart Gurney, Laboratory Director, Loss Prevention Department of Liberty Mutual Insurance Company, were published in *National Safety News*, October of 1940. The work of Liberty Mutual paralleled the research of Sidney V. James for Underwriters Laboratories, and both research quests led to the development of test instruments for a reliable means of evaluating flooring surfaces. The research of James was collaborated with field experience and laboratory measurements, which ultimately led to the development of the machine, named in his honor, which continues to be utilized today for measuring slip resistance of products used for maintaining walkway surfaces.

A unique feature to the James Machine was the establishment of a performance criteria value for the static coefficient of friction. James' research correlated the results of laboratory tests and tests made on actual floors under service conditions, as well as field experience for several years, that culminated in his recommendation of a minimum value as an acceptable coefficient of friction.

This value recommended by S. V. James in 1945 to the Casualty Council of Underwriters Laboratories, Inc., was judged to provide a surface with a sufficiently high static coefficient of friction to provide walkway surfaces presenting a reasonable risk. A static coefficient of friction of 0.5 or greater thus became the UL criteria for slip resistant polishes [3].[3]

ASTM Committee D 21 on Polishes formed a subcommittee for the continual investigation of slip restance in 1950. The developmental work of James was a cornerstone to the eventual standard test method for "Static Coefficient of Friction of Polish-Coated Floor Surfaces as Measured by the James Machine D 2047," which continues to be used today. ASTM D 2047 remains to date, the first and only voluntary consensus standard that specifies a compliance criterion, namely, 0.5 static coefficient of friction [4].

The development of a machine to evaluate slip resistance properties provides a means of objectively examining surfaces for potential use as flooring materials. The acceptance criteria of 0.5 SCOF as measured by the James Machine offers a basis for determining if a flooring surface and maintenance system may provide the sufficient traction properties for pedestrian traffic. Examining a selection of walkway surfaces and their resulting static coefficient of friction values allows a better understanding of the complexity of floor care, surface selection, and pedestrian safety. Although a surface rendering a consistent value

[3] A paper given at the 1971 CSPA Mid-Year meeting gave a theoretical justification for the 0.5 static coefficient of friction criteria for a safe walkway surface, based on biomechanical studies of human walking. They found that a static coefficient of friction of 0.3 should be sufficient for normal walking over a clean surface with leather-shod shoes. Ekkebus, C. F., and Killey W., "Validity of 0.5 Static Coefficient of Friction (James Machine) as a Measure of Safe Walkway Surfaces", *Test Methods and General Information*, January 1977 277-80 Chemical Specialties Manufacturers Association, Inc.

While reassuring, this theoretical analysis is not as important as the strength of the statistical evidence gained from the industry's experience with millions of gallons of floor finishes.

of 0.5 SCOF does not guarantee that accidents are not possible, it does provide for a reasonable risk for pedestrian walkways.

The Walkway Surface

Among the multitudes of flooring surfaces available today, the vast majority of those materials provide firm, stable, and slip resistant surfaces. Published traction values for a variety of unpolished flooring surfaces from Templer, Zimring, and Wineman, 1980, provided values that were listed as adapted from the British Standards Institution 1984, which gave ranges of values from 0.4 to <0.75 for linoleum tile, concrete, cast iron, terrazzo, vinyl asbestos tile, and clay tile. Values for carpet, textured clay tile, PVC, cork, rubber and Carborundum finish clay tile were listed as >0.75 [5]. The reference table did not specify the actual test procedure used for attaining the numbers listed, but they are within reasonable expectations from research and experiential evaluations on similar surfaces.

Slip resistant testing of surfaces used for flooring applications has provided interesting observations regarding the interaction of those surfaces and the resulting traction that can be expected. An early report issued by Underwriters Laboratories to Hillyard Chemical Company on February 28, 1947, showed the effect of a variety of cleaning compounds and maintenance materials on asphalt tile, rubber tile, linoleum, and quarry tile [6]. Those results showed the use of typical maintenance items in use at that time would maintain the surface traction level of 0.5 or higher.

The reported findings by UL indicated slip resistance of an untreated flooring surface that may not provide a static coefficient of friction value of 0.5, could be altered through the use of maintenance products to increase the friction to the 0.5 level and beyond. Depending upon the original value for that surface, the measured slip resistance could be enhanced to varying degrees. Comparing the untreated surfaces to those with a maintenance coating applied, the resulting static coefficient of friction value may be slightly reduced or increased, and in nearly all instances would exceed the 0.5 value established by Underwriters Laboratories and listed by ASTM D 2047.

Among the factors influencing the traction available for a flooring substrate is the surface profile, as well as the contact area allowing the pedestrian's foot to mate adequately with the flooring surface. In extremely irregular flooring surfaces, the profile of the floor may be too erratic to allow for a consistent mating of the pedestrian's foot and the floor. In those situations, providing a more consistent flooring surface through the use of a floor polish can provide a uniform surface for traction. As layers of floor polish film are formed across the floor, the traction of the floor will shift from the flooring surface to the polish applied to the floor. This observation applies to not only the polish film, but also to the maintenance products intended for use on the flooring surface. The evaluation of ancillary items used for floor maintenance also exhibits the similar trend of complementing the traction properties of the floor surface, based upon their individual composition.

Resilient flooring surfaces can be altered through surface roughness, or through polish applications to increase the surface contact area with the pedestrian's footwear. In the case of burnishing or buffing floor polishes on resilient flooring, the grinding process smoothes the surface, thereby increasing the contacting surface areas. A floor that has been burnished or buffed will typically increase its aesthetic value in appearance and also maintain or increase its perceived traction. This process can be confirmed through evaluating coated panels that are designed to be removable for research purposes, or through direct evaluation.

The selected panels can be treated in a variety of maintenance operations and then removed for slip resistance testing on a James Machine, or subjected to "in-situ" testing by a number of methods, including walking across the floor and comparing the resulting traction feel. Although this method can be taken as a subjective evaluation, an experienced floor maintenance practitioner can detect subtle differences between surfaces that can be confirmed through laboratory testing. This procedure was formalized through the Chemical Specialties Products Association as Bulletin 245-70.[4] This process has been used in practice by flooring companies for years and a variation of this procedure is described as a guide for evaluating floor polishes under ASTM Standard Practice for Rating Water-Emulsion Floor Polishes (D 3052-87) through ASTM D 21 committee on polishes. This method of evaluating flooring products builds upon the concepts used by the flooring industry and was a principle concept in the research by S. V. James for UL.

Back to Basics

Examining flooring surfaces under real work and laboratory research has provided a wealth of insight into the expected traction potential properties of these surfaces. A relatively common design feature of office building construction may incorporate form and function through the use of granite in a variety of surface profiles. The exterior of a building may begin with the use of flame cut or highly textured granite flooring. The textured surface profile of flame cut granite allows for anticipated inclement weather to channel from the surface and minimize hydroplaning from foot traffic. The same material can then transition to a honed surface as the granite walkway surface is brought into the building. Once inside, the honed surface may become polished throughout or as accent areas and, in the process, provide good traction that is aesthetically pleasing and provide relatively low maintenance requirements to the building.

The use of resilient flooring provides a variety of interior flooring options popular for educational, institutional, and commercial settings. A patterned or smooth surface provides options for the building occupants depending upon their individual needs. The use of maintenance coatings (waxes or polishes) prolongs the life of these floors, and adds value in appearance, performance and ease of maintenance, as well as providing a firm, stable, and slip resistant surface. The application of a floor polish will fill the microscopic surface irregularities with a versatile protective polymer film. In this spirit, the polish will equalize the slip resistance properties across the entire profile of the floor, due to its taking on the characteristics of that polish film. The use of these maintenance polishes allows a sacrificial film to form, protecting the flooring surface from abrasion and wear, while providing a uniform traction profile across the entire floor.

[4] Consumer Specialty Products Association - 1913 Eye Street NW, Washington, DC, 20006.

Heavily textured resilient floors may require additional maintenance needs due to their surface profile. Sharp peaks can become smooth through traffic and wear, and due to the more aggressive maintenance needs of these floors, they can eventually become filled with soil and debris. The use of a floor polish in these situations, as long as their use is in a dry environment, can lessen the overall maintenance requirement as the surface becomes smooth. Provided the surface roughness was not a specific design feature for a specific use, a floor polish will ease the maintenance requirement of the floor while providing a uniform level of slip resistance.

An interesting feature of resilient flooring is the viscoelastic properties of these surfaces under traffic that allow the surface to conform slightly to the pedestrian's foot. As an individual walks across the floor and the contact areas meet, the more those contacting surfaces are squeezed by the normal (downward force), the greater a contact area is formed, resulting in more potential friction between those surfaces. The compression of the floor is typically only momentary as the resilient floor returns to its original shape.

Although the amount of friction between two dry surfaces does not depend upon the contact area, (17th Century French physicist Guillaume Amontons), if the viscoelastic properties of the two mating surfaces is allowed to increase, the macroscopic friction is enhanced by the adhesive forces of those two surfaces to one another. Even hard surfaces (diamond) can become somewhat plastic under room temperature, provided sufficient pressure is applied [7].

In regard to rough surface textures, the floors' surface roughness may play a role in "stick-slip" friction, in which surfaces gliding past one another momentarily cling and then let go [8]. In the case of highly textured floors or footwear, surface roughness or contact roughness can impart a negative effect on the floor and increase the potential of a trip and fall, or stumble. This property is one that should be addressed by the building design, depending upon the intended use of the floor and facility. An example of this potential is in the use of viscoelastic surfaces for athletic events. If the surface is too easily compressed for the activity placed on that surface, there can be problems, i.e., basketball resulted in more injuries to the lower extremities than for alternative surfaces due, in part, to an excess of surface friction [9].

Other Considerations

The flooring surface is an important factor in the evaluation of walkway safety, but it is only one of several influences to the overall success of a facility maintenance plan. The introduction of contaminants onto the floor, whether they are wet or dry, temporary or ongoing, needs to be considered. The effects of hydroplaning on a smooth surface from wet contaminants can greatly affect the available traction of the flooring surface. Controlling and minimizing contaminants, (water, snow, ice, spills, debris, etc.) will prevent a temporary compromise of the slip resistance of the floor. Hydroplaning over a liquid spill on a smooth floor can quickly reduce the friction of the floor and lead to a fall. The hydrodynamics of the liquid spilled upon the floor will determine, to a degree, the extent of the change in surface friction, along with the planarity of the surface and composition of the floor coating. Due to the complexities of hydroplaning on the floor, it is best left as a future topic of discussion.

Dry contaminants can also affect the available surface friction of the floor, which is influenced by the hardness of the flooring substrate and the nature of the dry compound. Construction dust, for example, will potentially act as miniature roller bearings on a hard flooring surface, thereby reducing the traction. Again, controlling the source of contaminants and removing these types of dry lubricating substances will restore the friction of the floor.

The type of footwear of the pedestrian also plays a role in the traction that is measured as well as perceived by the individual using the floor. The work of S. V. James was developed around the use of leather-soled shoes. Differences in footwear patterns have been studied by a number of organizations in attempts to develop the ultimate shoe. A paper presented at an ASTM symposium in Denver, Colorado, in 1989 included the statement, "Under some conditions slip is unavoidable. Likewise, no single sole design will be 'best' in all conditions" [10]. Whether the footwear contains a hard or elastic sole, the pattern design in the sole, and the condition of the footwear will all contribute to the perceived traction available to the pedestrian. A major manufacturer in sportswear used an interesting concept in footwear sole material in a specialized sole, which was designed to mimic the feet of mountain goats. The shoe outsole had soft, goat-like pads that compress, which allowed the harder, outsole rim of the sole to bite into the ground surface, allowing the wearer to commit to each foot plant on rough terrain.

Summary

The type of flooring surface and selected maintenance process can impact the resulting traction of the chosen pedestrian walkway. The surface structure, whether it is resilient, mineral, wood, or other hard surface type, can be safely maintained for pedestrian traffic, provided quality products and a little common sense is applied. It is a rare situation to find a dry floor that exhibits questionable frictional properties; however, a contaminated floor can create a reduction in available traction and those instances should be avoided, or at least minimized. The use of commercial polishes for the care and maintenance of both resilient and other flooring surfaces protects and enhances the substrate for a longer service life, while maintaining the surface friction for pedestrian use. A little common sense and good housekeeping practices will provide a firm, stable, and slip resistance walkway surface.

References

[1] James, S. V., "What is a Safe Floor Finish?," *Soap and Sanitary Chemicals*, October 1944.

[2] Gurney, S., "Is the Floor Too Slippery?" *National Safety News*, August 1940, Vol. 22-23, p. 63.

[3] Ekkebus, C. F., and Killey, W., "Validity of 0.5 Static Coefficient of Friction (James Machine) as a Measure of Safe Walkway Surfaces," *Test Methods and General Information*, Chemical Specialties Manufacturers Association, Inc., Washington, DC, January 1977, pp. 277-280.

[4] Sacher, A., "The Application of Forensic Biomechanics to the Resolution of Unwitnessed Falling Accidents," *Journal of Forensic Science*, September 1996, pp. 776-81.

[5] Templer, J., *The Staircase Studies of Hazards, Falls, and Safer Designs*, The MIT Press, Cambridge, MA, 1992.

[6] Sacher, A., "Slip Resistance and the James Machine 0.5 Static Coefficient of Friction – Sine Qua Non," *Standardization News,* August 1993, pp. 52-59.

[7] Hayward, I. P. and Field, J.E., "Friction and Wear of Diamond," *IMECHE*, 1987 C159/87; pp. 205-209.

[8] Krim, J., "Friction at the Atomic Scale," *Scientific American,* October 1996; pp. 74-80.

[9] Maple Flooring Manufacturers' Association, "A Summary of Findings, Incidence of Injury Study: Maple Flooring vs. Synthetic," 1997.

[10] Wilson, M.P., "Development of SATRA Slip Test and Tread Pattern Design Guidelines," ASTM STP 1103, pp. 113-123.

Theodore Tysak[1]

Polish Maintenance for Fun and Profit

Reference: Tysak, T., **"Polish Maintenance for Fun and Profit,"** *Technology of Floor Maintenance and Current Trends, ASTM STP 1448,* W. J. Schalitz, Ed., ASTM International, West Conshohocken, PA, 2004.

Abstract: The cost of polish maintenance is many times greater than the cost of the chemicals and their application. The physics and chemistry of the various procedures and processes in polish maintenance are discussed, with an emphasis on high speed burnishing.

Keywords: crosslinking, dynamic mechanical analysis, electron spectroscopy for chemical analysis, formulating, gloss, high speed burnishing, maintenance, machine maintenance, spray-buffing, tensile loss modulus, time-of-flight-secondary ion mass spectrometry

Introduction

For those familiar with performing floor maintenance, it does not take long to realize that it is a costly process. The cost of labor is an expensive component of floor maintenance in today's market environment, and it takes ample work to keep a floor in peak condition. Floor maintenance procedures including cleaning, buffing, stripping, and applying a finish all require considerable labor input.

The impact of labor in floor care has triggered continuing attention to "extending" the money available for maintenance. Efforts in this direction have taken several forms. One of the most important has been the movement to more durable finishes. Greater durability equates to less frequent maintenance. If one polish lasts longer than another, the floor covered by the more durable coating needs less labor input because stripping and recoating would be reduced.

The demand for greater durability with a focus on reducing the labor associated with floor maintenance has provided the momentum to polymer manufacturers and polish formulators to design new products to meet this need.

---

[1]Principal Scientist, Floor Care, Rohm and Haas Company, 727 Norristown Road, Spring House, PA, 19477.

Enhanced products denote only one aspect of the floor care industry's cost-management activities. Skilled practitioners in the field have also sought means to make the most of high-performance polishes based on the new vehicles that push the durability envelop to their fullest advantage. This interest has led to different approaches in floor maintenance. The introduction of zinc crosslinking to floor polish binders in the mid-1960s significantly improved the traditional mop and recoat procedures used during that time period. The use of rotary buffing machines to refurbish "worn" polish complemented mop and recoat procedures. Similarly, with zinc moving the market in a new direction, the introduction of high speed burnishing was the next influential process that changed floor care maintenance and moved toward reducing maintenance costs. "Machine maintenance" offers the option of an attractive floor finish for less time, effort, and money than simply relying on recoating the floor periodically. When machine maintenance is combined with a polish based on an outstanding vehicle and formulation, the floor care professional can achieve significant floor appearance at reasonable costs.

Floor Maintenance Techniques

The first part of any floor maintenance procedure is the stripping or removal of old finish from the floor. This process is accomplished by exposing the floor to aqueous amine/solvent stripping compositions and abrasive rotating fiber pads. Care must be taken to use proper stripping chemicals and procedures designed for floor finish removal. Electron micrographs (Figure 1) show the damage caused by excessive scrubbing with a high concentration stripper. For comparison, an electron photomicrograph of the same stripper, but with water dilution in place of the solvent, 2-butoxyethanol, typically present in stripper formulations, is also shown. The tile damage (seen as tile dissolution) and increased porosity is obvious. Although this damage is not apparent early in the life of a flooring substrate, repeated or excessive stripping will cause the flooring substrate to have a reduced "useful" life expectancy.

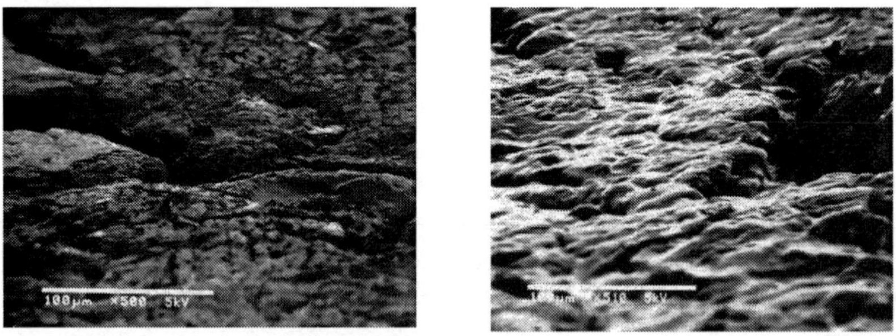

Figure 1 - *Scanning Electron Micrograph, right, (500 X magnification) shows the damage caused by scrubbing (400 cycles, Gardener Straight-line Scrub Machine, 3M Green Scrub Pad) with a concentrated high performance stripper containing >30% 2-butoxyethanol. For comparison, a scanning electron micrograph of the same tile, scrubbed identically with the same stripper but with water in place of the 2-butoxyethanol., left.*

Following the stripping operation, application of four to six (or more) coats of the base polish occurs. The polish film is very thin and typically measures a few thousanths of an inch in thickness (Figure 2). Afterward, usual procedure calls for the polish-covered floor to be swept, mopped or auto scrubbed on a daily basis to remove dirt and soil that has been tracked onto it. The constant exposure to pedestrian traffic and frequent cleanings combine to damage the surface of the polish and cause it to wear as time passes. The surface becomes rough, which detracts from its gloss and general appearance over time. To refurbish the floor coating's appearance, a new layer of finish must be placed over the "damaged" layer(s).

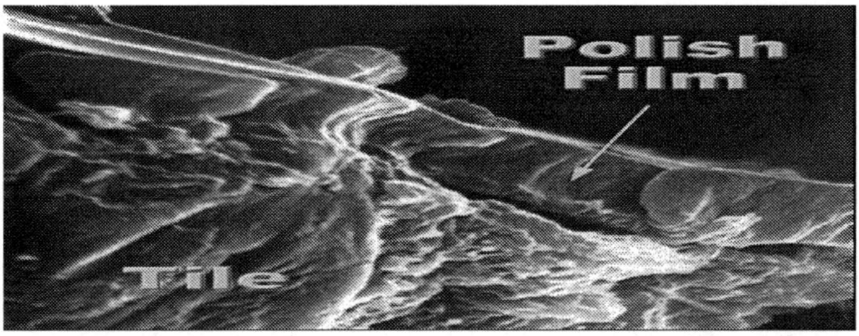

Figure 2 - *Photomicrograph of polish film (25% polish solids, after four coats applied with cotton blend mop) on vinyl composition tile. Film thickness was determined to be approximately 25 microns.*

Because traffic patterns are never uniform, the polish usually deteriorates irregularly. In some areas, damage is so heavy the polish film may be completely removed from the floor. In other places, the floor coating may be virtually unmarked. Polish applied over a worn layer of finish will exhibit some of the subsurface irregularities and vary in film thickness. The resulting lack of smoothness and uniformity in the new material detracts from its gloss. Furthermore, the base polish will eventually degrade so badly that a new coat cannot improve its appearance. At that point, it must be stripped completely and new polish applied.

Gloss or "wet look gloss" can be obtained from any floor polish simply by applying a sufficiently heavy, multi-coat loading of polish to cancel out completely all of the microscopic contours and features of the substrate (Figure 3). Machine maintenance differs from basic recoat processes in that it does not rely essentially on the application of new polish to bring back the appearance of the floor. Instead it utilizes a rotary buffing/burnishing machine to scour or grind away the surface irregularities and damage (Figure 4) that have developed in the polish surface. In essence, the machine removes a small fraction of the damaged polish film to expose an unblemished layer of polish. Because the burnishing action created a smoother surface, the polish possesses the sheen comparable to freshly applied floor coatings.

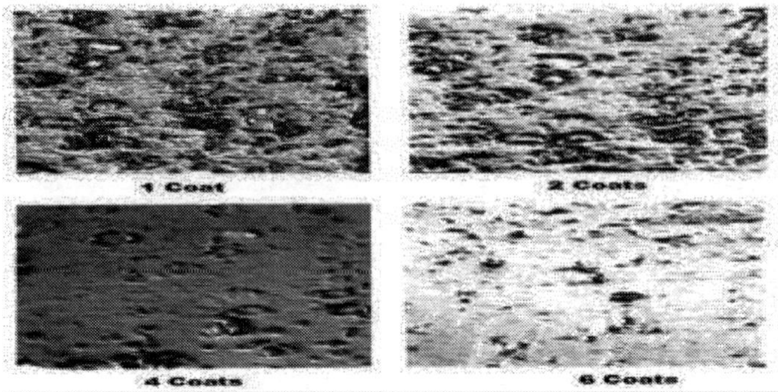

Figure 3 - *Photomicrographs (100 x magnification) showing film of conventional polish applied to vinyl composition tile. As the number of coats increases from left to right, the appearance of the surface becomes smoother.*

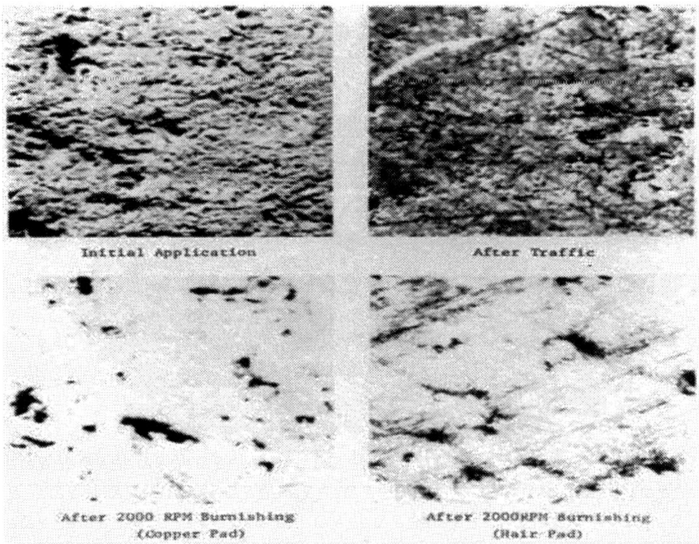

Figure 4 - *Photomicrographs (100 x magnification) showing floor polish performance before and after machine maintenance. Note that in this example, the finish is better suited for a synthetic pad (copper pad) with less abrasive qualities because of the lack of surface scratches evident from the photomicrograph.*

With machine maintenance, the contractor may apply new polish more frequently in some cases rather than simply relying on non-machine maintenance recoat procedures.

This is because the constant buffing or burnishing removes layers of the polish. New coats of finish must be applied to maintain polish film thickness. However, the constant machining also prevents the polish buildup that characterizes floors that rely solely on non-machine maintenance recoat procedures. Since a maintenance contractor does not have a significant buildup of polish, the contractor can strip and recoat the floor less often for a savings in labor. Even if a floor care professional must apply polish more often with a machine maintenance program, the reduced frequency of stripping and recoating compared to a non-machine maintenance program is a benefit. The constant buffing and burnishing also guarantees that there is never as much fall-off in appearance as with floors relying on non-machine maintenance methods.

Applications of sufficient base polish is the cornerstone of any machine maintenance program. Depending on traffic conditions, machine maintenance can start the day following polish application or up to a week later. With a spray-buffing regimen, a liquid medium is applied to the floor, usually in the form of a spray at the front of the machine. Due to its composition, the spray-buff medium swells the polish film, lubricating it to make it more amenable to the buffing procedure. The rotary floor machine is then used to buff the finish while in a wet state. This process is multi-functional, both cleaning and repairing the polish surface. The initial gloss and clarity of the floor coating is restored as the spray-buff medium dries during the buffing operation due to the friction generated between the pad and the polish.

The progression of events in burnishing is similar to the agenda for spray-buffing but with several key differences. Burnishing is usually done to a dry floor or to one that has been coated recently with a "dressing" or pre-burnish dressing. In addition, burnishing requires very high speed (>1000 rpm) floor machines and different pads than spray-buffing.

Graphs I and II, depicted in Figure 5, demonstrate the advantages of machine maintenance (spray buffing). As the first graph shows, polish applied and maintained without machining starts off with an excellent appearance rating and then deteriorates. Immediately after application, the polish will register 9 on a 10-point scale of overall appearance. (Appearance ratings are based on a combination of gloss, gloss retention, and the relative absence of scuffs, soil, and marks.) Note, however, that appearance decays rapidly under traffic. After some time has passed, the rating has fallen below 3. A new coating of polish must be applied to restore the floor to its original appearance.

The second graph shows the behavior of a floor finish designed for spray-buff maintenance. The solid line indicates its performance when maintained without buffing. Note that the finish exhibits the same pattern of deterioration as the finish in Graph I. The saw tooth line in Graph II shows what happens to the polish appearance when spray-buffed. After one week on the floor—before it has been buffed the first time—the finish shows the same deterioration as the non-spray buffed finish. However, the initial buffing dramatically improves the coatings appearance and the gloss becomes almost as good as when the finish was new.

In between ensuing buffing sessions, the polish shows the same pattern of decline as the polish maintained by non-machine maintenance procedures. The weekly spray-buffing, however, actually boosts luster above the original level by the end of the fourth week. The improvement in the appearance of the spray-buffed formulation contrasts with the continued decline in the shine of the non-machined floor coating. Each week the appearance gap between the two polishes grows wider. Before the polish is spray-buffed at the end of the fourth week, it registers better than 8 on an appearance scale compared to the less than 3 of its "non-machined" counterpart.

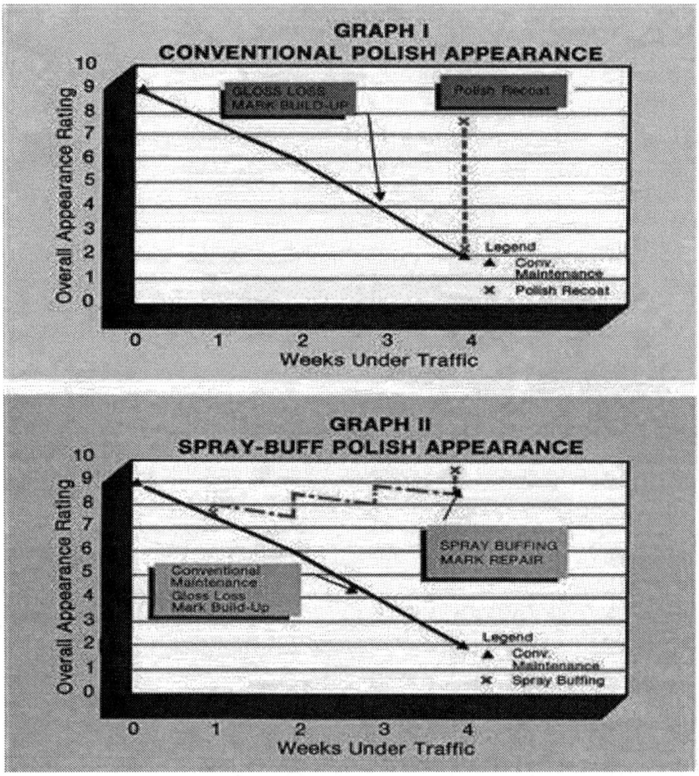

Figure 5 - *Conventional polish maintenance represents mop and recoat maintenance procedures that do not rely on machine restoration processes.*

One should note that burnishing would generate the same pattern of appearance and wear as spray-buffing although the specific ratings would be different. Consequently, the broken line in Graph II is also representative of burnish maintenance.

Spray-Buffing Systems

Vital to any spray-buff floor care program is the rotary buffing unit. There are, however, three other equally important elements in this type of maintenance: buffing pad, base polish, and spray medium. The four items are interdependent; consequently, varying any one can dramatically alter the finish's gloss, overall appearance, and performance. The maintenance contractor must employ a "systems approach" to create a spray-buffing program. The function of each element is critical and yet is related to the others.

Floor Machines

Equipment used to rotate the buffing pad on top of the polish surface represents the first element of the spray-buffing system. Many different machine models are available, however, all function in approximately the same manner. The most important characteristics from the standpoint of the spray-buff system are the speed at which the machine drives the pad and the weight of the machine on the pad.

Buffing machines can be classified into generally four categories on the basis of rotational speed of the disk/pad assembly . Low-speed machines operate below 300 rpm; medium-speed machines in the range of 300 – 700 rpm; high speed units at better than 1000 rpm; and ultra high speed floor machines at speeds of 2000 rpm and higher.

The differences in operating speed affect both application process and the final maintenance results. In low- and medium-speed units the full weight of the machine typically rests on the attached floor pad. Conventional practice requires the operator to swing the machine from side to side as the operator moves forward; some skill is necessary to use and manipulate the buffer in a manner that will result in a favorable outcome. To obtain optimum results, the user must make several passes over an area with a low-speed machine and usually, only half as many passes with a medium-speed buffer. The result is that a person can buff much more floor space in given time than with a slower machine.

In contrast to slower units, the high-speed machines sit on wheels that bear most of their weight. The pad is in contact with the floor finish, but it does not have the full weight of the machine pressing down on it. These units move only in a straight line and are considered straightforward to handle with minimal instruction. Because this type of machine rotates the pad so much more rapidly than the lower speed-models, often only one pass over the floor is required. Moreover, the floor machine can be moved faster so each pass takes less time. Consequently, manufacturers claim that high-speed equipment can cover more square footage per hour than the slower, low and medium speed machines.

The properties of gloss build and the capacity to repair are both influenced by a machine's pad operating speed and the weight the pad imposes on the floor. Gloss enhancement normally improves as machine speed increases, but the ability of the polish to be repaired generally diminishes. The decrease in repairability is due to the fact that the contact of the pad on the floor is less. Consequently, the pad does not grind as deeply into the surface of the polish as do the pads on the slower units. The end result is that high-speed units are not efficient at removing deep scuffs and marks from the polish film. On excessively scuffed polishes, high-speed spray buffing repairs very little and may not produce an acceptable surface.

Floor Buffing Pads

The second major component in the spray-buffing system is the buffing pad. It has three purposes: to clean, repair, and polish the finish.

Pads used for spray-buffing purposes are usually made from polyester or nylon fibers. Spray buffing pads are variants of pads employed to strip polish from a floor; they differ in one prominent respect. Spray-buff pads abrade less and polish more than the stripping pads. Most pad suppliers differentiate between pad types by using a color-coding system. Generally, the color red is used to designate a spray-buff pad; black pads are employed for stripping. Manufacturers offer other types of synthetic pads for use with high-speed (> 1000 rpm) machines. A number of these products have had the fibers modified so that contact friction between the floor polish coating and pad increases. High speed machines equipped with these modified pads exhibit outstanding polishing strength.

Base Polish

The base polish employed in a spray-buff system needs a wide range of performance properties. Obviously, it must possess the characteristics required of any floor finish: high gloss, leveling and wetting properties for a smooth surface, and the ability to form a tough, durable film. Further, spray-buffing demands that the base polish form a uniformly good coating, with sufficient film integrity that typically applied coats can withstand the abrasive forces of buffing. It must be clear upon lay down and resist yellowing and hazing over time. Good color retention is a must, given the extended time between application and stripping. Spray-buffing base finishes should also be very resistant to soiling and marking, be difficult to scuff, and tend not to powder. With repeated buffing, the base spray buff polishes should be able to retain their slip resistance. These polishes should be easy to remove when it finally comes time to strip them. Last, they should be compatible with other elements of the spray-buff system so that no hazing occurs.

Spray-Medium

The last major element of spray-buffing systems is the medium. Formulations used as buffing media are applied in a fine stream or mist to the trafficked polish surface just before buffing. The role of the medium is to assist in cleaning the surface of the base polish, to lubricate it against the abrasive effects of the buffing pad, to hold the abraded soil and powder to the pad, and to maintain key properties (e.g., slip resistance) of the existing finish.

In addition to these properties, the spray-buff medium formulation must be easy to remove from the buffing pad. Removability is important because the polish ground off in the buffing process becomes suspended in the medium, having a "gummy" consistency. As has been noted, a good spray-buff medium retains the dirt, dust, and debris in the buffing pad. If the medium is difficult to remove from the pad, the contractor must either spend valuable time and effort removing it or replace pads frequently. Both options can be expensive.

Historically, floor finishers employed floor polish typically diluted to 2.0 – 6.0% solids as the spray medium. Dilute floor polish did an adequate job of cleaning and lubricating, and it had the additional benefit of being readily available to both formulators and maintenance personnel. The use of dilute floor finish as a spray buff medium continues by some floor care professionals. New products, however, based on modifications to the "dilute" floor finish formulations have been made available to the marketplace and are improved alternatives.

Formulating for Spray-Buffing Applications

As indicated previously, spray-buffing places an arduous demand on base polishes and spray media. Consequently, the best products employed in both these roles are high-quality conventional polishes based on metal crosslinked (zinc) aqueous acrylic polymers. Polishes based on today's metal crosslinked vehicles posses the right blend of gloss, clarity, toughness, and durability for this kind of floor care regimen.

Commercial experience has confirmed that standard acrylic floor polish formulations are acceptable in spray-buff applications. However, the industry has found that specially designed base polishes and media deliver improved response to spray-buffing without sacrificing other performance properties. The new finishes are derived by making relatively minor adjustments in the composition of existing floor coating formulations, such as to the Polymer/alkali-soluble resin(ASR)/Wax ratio.

Base Polishes

The first key to maximizing spray-buff response is employing the right vehicle in the base polish. The polymer is responsible for the durability, gloss, and, removability of the finish. Standards for performance in these areas have been established by the use of acrylic compositions relying on metal crosslinking with the correct choice of aqueous based polyethylene and polypropylene wax emulsions. Choosing the proper raw materials is only the first step in formulating a "spray-buffable" base polish. The formulator must also blend the ingredients in the right proportions. Generally this necessitates adjusting the ratio of polymer, alkali-soluble resin (ASR), and wax. High polymer and low wax levels tend to provide polishes with optimal spray-buff performance.

High Speed Burnishing

As one may have concluded thus far, gloss is a surface phenomenon. The gloss of a floor polish film depends on the nature of the polish film surface. There is a compositional, or a chemical element to gloss: polymers containing styrene produce higher gloss floor polish films than their non-styrene-containing analogues. Styrene contributes to the bulk refractive index of the coating film and to film forming characteristics that result in smoother films. It stands to reason that smoother films produce glossier films. Multiple coat applications of floor polish will result in successive reduction of the substrate surface irregularities at the polish surface, so that with each application of a coat of polish, the resulting polish film surface becomes smoother and glossier (Figure 3). As the polish film surface becomes smoother in multi-coat applications of polish, it approaches the regularity or uniformity of reflection, and clarity of reflection that is usually associated with a wet surface. This smooth, planar surface has a "wet look," or wet gloss. The roughness and porosity of the underlying flooring substrate, and the depth of the polish film, have no effect on this perception of polish gloss, if the polish film surface is sufficiently smooth.

As has been demonstrated previously (Figure 3), high gloss or "wet look" gloss performance can be obtained from any floor polish simply by applying a sufficiently heavy multicoat loading of polish to completely reduce the "telegraphed" surface out of all of the minuscule contours and features of the substrate.

High speed burnishing provides improved gloss for a floor polish by abrasively smoothing the polish film surface (Figure 6). This smoothing process entails the removal of the polish surface imperfections, producing a smoother, glossier film surface. The polish toughness and abrasion resistance, the aggressiveness of the pad, and the dynamics of the high speed machine affect the ease with which this smoothing process takes place.

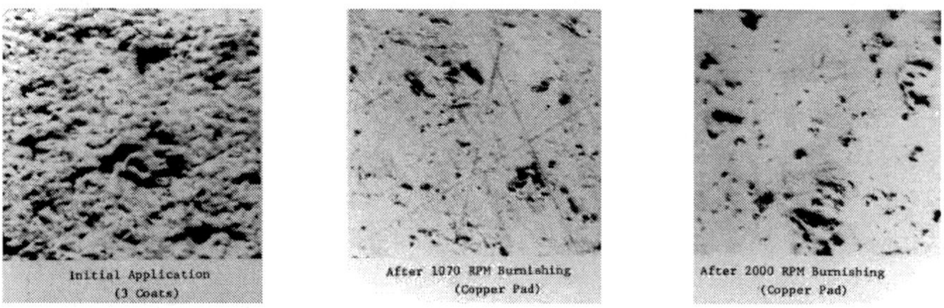

Figure 6 - *Floor Finish performance and relationship to machine speed. The pad is the same, however, the performance differs by increasing the pad speed as noted in photomicrographs: little to no scratches from the synthetic (copper) pad present in the finish after 2000 rpm burnishing. In each case the burnishing machine was electric powered.*

The floor polish film is not moved, or caused to flow by high speed burnishing. Figure 7 shows tensile loss modulus data for six current, commercial floor polishes. It is evident from the data generated that film deformation is apparent at temperature levels exceeding 75 °C. Similarly, the "work function" or "Tan δ" (Figure 8), which is another method to assess mechanical stress applied to amorphous films (i.e., floor polish coatings), indicates that energy absorption and dissipation of floor polish films occur at temperatures exceeding 100 °C. Dynamic mechanical analysis of polish films show that heat will affect polish films, however, the residence time of the burnisher is not sufficient to cause polymer flow. Indeed, all the evidence confirms that burnishing is like spray-buffing, essentially an abrasive, grinding process. The burnisher abrades away the thin, top layer of polish containing some of the dirt, scuffs, and the marks that have adversely affected gloss and appearance. Additionally, transmission electron micrographs (Figure 9) of a polish film exposed to high speed burnishing (2000 rpm propane driven burnishing with natural fiber pad) indicates that there is no differences between the bulk film appearance before and after the machine maintenance process.

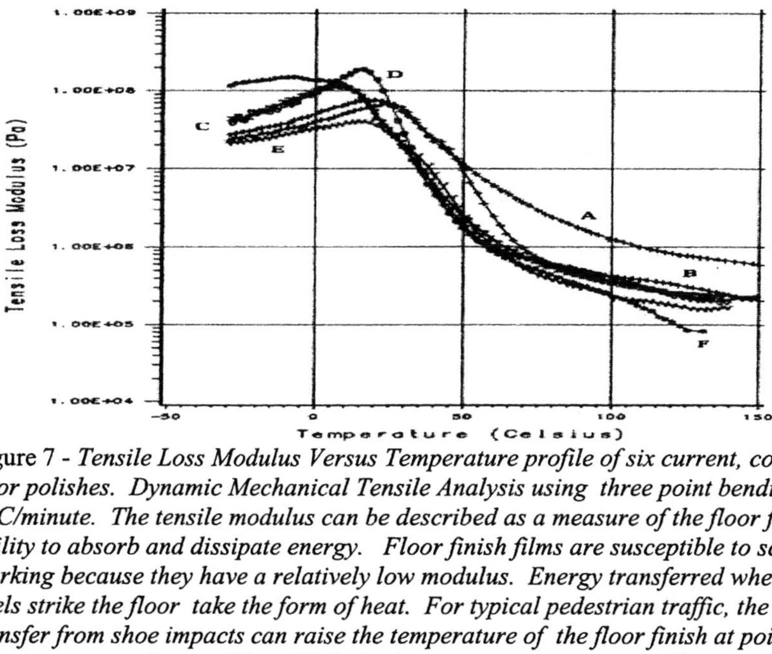

Figure 7 - *Tensile Loss Modulus Versus Temperature profile of six current, commercial floor polishes. Dynamic Mechanical Tensile Analysis using three point bending at 1 Hz, 4 °C/minute. The tensile modulus can be described as a measure of the floor finish's ability to absorb and dissipate energy. Floor finish films are susceptible to scuffing and marking because they have a relatively low modulus. Energy transferred when shoe heels strike the floor take the form of heat. For typical pedestrian traffic, the energy transfer from shoe impacts can raise the temperature of the floor finish at point of contact to range from 44°C – 60° C. In this temperature range, finishes tend to soften slightly and become susceptible to scuffing and marking. Obviously, finishes that possess a higher modulus (Finishes designated 'E' and 'F') will be less prone to scuffing and marking. In wear testing, with no maintenance, Finishes 'E' and 'F' performed better than finishes 'A' and 'B' in terms of black heel and scuff mark resistance. The other finishes, 'C' and 'D' were rated as better than 'A' and 'B' but not as good as 'E' and 'F'.*

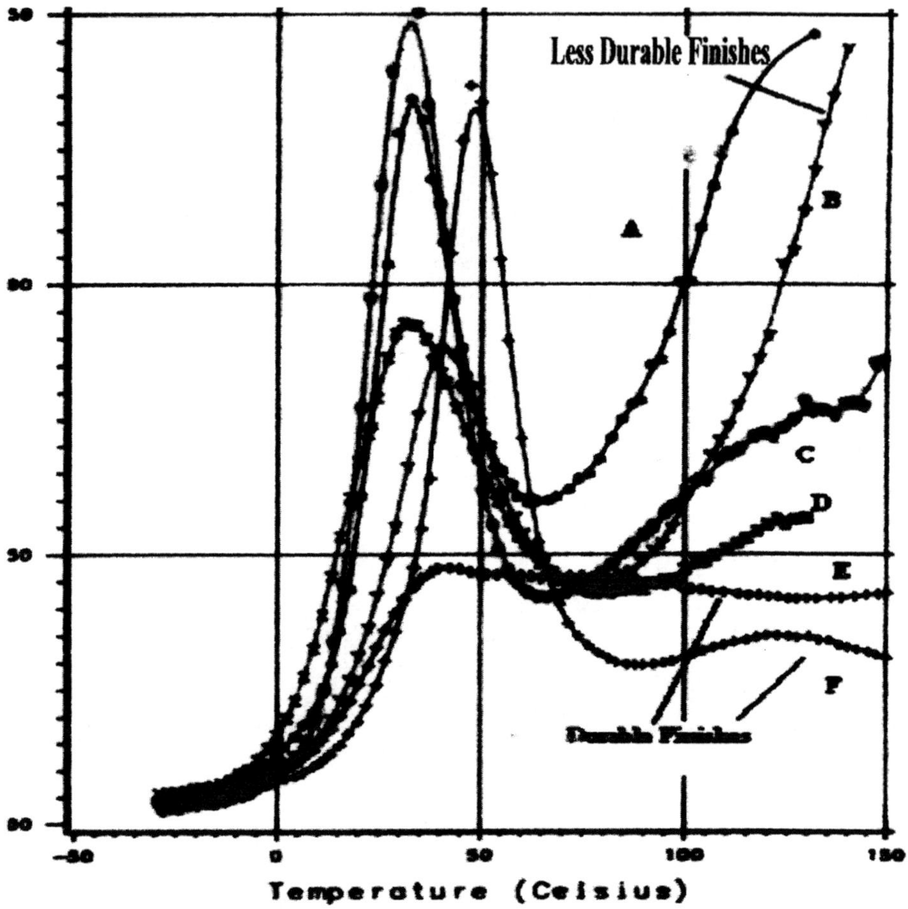

Figure 8 - *Dynamic Mechanical Analysis (Tan Δ versus Temperature) of six current, commercial floor finish products (three point bending at 1 Hz and 4 °C/minute). Note that temperatures in excess of 75 C are required to observe energy absorption/dissipation effects. Some finishes show no slope or a negative slope. The finishes exhibiting no slope or negative slope values (E and F) can absorb and dissipate more energy than the less durable finishes (A and B) and hence are capable of lasting longer on the floor. In floor testing, the "durable" finishes lasted twice as long as the "less durable" finishes before recoating was required. All of the finishes were tested in a food super market environment and maintained by daily scrubbing with a neutral cleaner and burnishing four times/week with a 2000 rpm propane driven machine.*

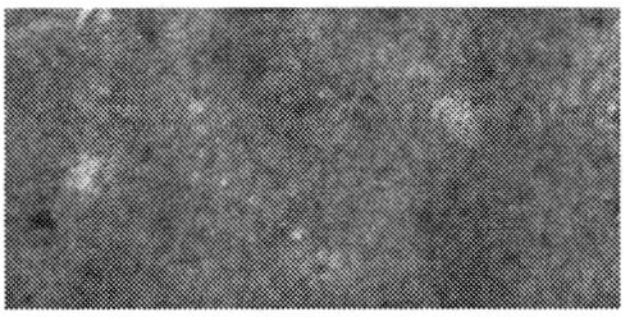

Micrograph of Bulk Film Before High Speed Burnishing

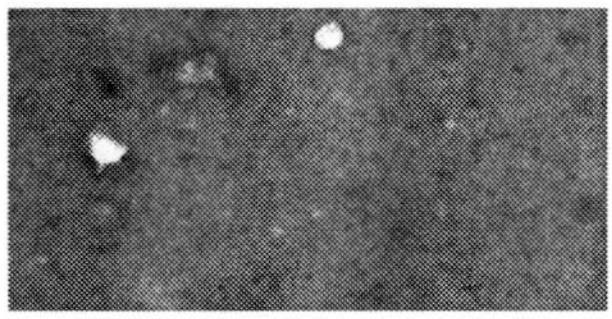

Micrograph of Bulk Film After High Speed Burnishing

Figure 9 - *Transmission electron micrographs of polish films before and after high speed burnishing. A propane machine rated at 2000 rpm fitted with natural fiber pad was used over a vinyl composition tile coated with four coats of a 25% solids floor finish (coating rate was 2000 square feet/gallon).*

High speed burnishing is not an efficient cleaning operation (by itself). Dirt and grime can be forced into the film surface by an unclean pad. Those surface soils that are not driven into the film will be projected into the air and add to the dusting normally present with high speed burnishing (an aggravated case of dusting.) The only way to mitigate these problems is to clean the floor thoroughly before high speed burnishing.

The cleaning operation will not only remove soils, it will also enhance the high speed burnish response. The materials (particularly the water) in a conventional cleaner formulation will serve to soften the surface of the polish film, making it somewhat less abrasion resistant. This softening effect lasts for 10 – 15 minutes after damp mopping, (and scrubbing) and is lost with the evaporation (or removal via vacuuming with auto scrubber) of the last remnants of water from the film. The cleaning operation serves to remove lightly imbedded soils, but it also improves the repair of traffic damage and promotes rejuvenation of gloss in the polish surface by temporarily reducing the polish abrasion resistance.

Detergent cleaning of the floor does change the surface characteristics of the polish. Time of Flight Secondary Ion Mass Spectrometry, TOFS-SIMS (Figure 10) and Electron Spectroscopy for Chemical Analysis, ESCA (Table 1) analyses reveal that only after several cleanings does the surface composition changes dramatically. Reductions in key elements (from ingredients added during the polish formulation stage) diminish. In the case of the TOFS-SIMS analysis, compositional characteristics of the film were monitored by focusing on a specific mass range that tracked the changes of the polyethylene wax component of the finish. Note that after repeated scrubbing over a 30 day period, the components normally associated with the polyethylene wax (a key ingredient of the polish) were not present at the surface. ESCA analysis used the polyethylene wax component as one of the monitored components; other elements were also studied. The data provided in Figure 10 and Table 1 was obtained by treating vinyl composition tiles with a commercially available neutral cleaner and employing an auto scrubbing floor machine fitted with a "blue" synthetic pad. As the data indicates, the floor finish is "attacked" and care must be taken to follow label directions for proper use of the cleaner. High concentration of cleaner or the use of stronger, alkaline cleaners will cause film degradation to accelerate. Although a requirement, the constant cleaning process leads to film degradation and combined with the abrasive process of the high speed burnishing procedure, contribute to the need for polish surface restoration in the form of a deep scrub and recoat or total polish film stripping and new coat application. Because the cleaning operation momentarily lowers abrasion resistance, the polish surface will be more readily abraded away.

As was mentioned for spray buffing, topical applications of aqueous based material can also be performed with high speed burnishing. As with spray buff mediums, a high speed restorer is a formulation that permanently softens the polish surface to augment the abrasive smoothing and polish film surface removal processes. A restorer improves the high speed burnishing repair by decreasing the polish surface abrasion resistance so that the film can be quickly abraded down to the level of surface defects caused by traffic.

Positive Ion SIMS Spectra

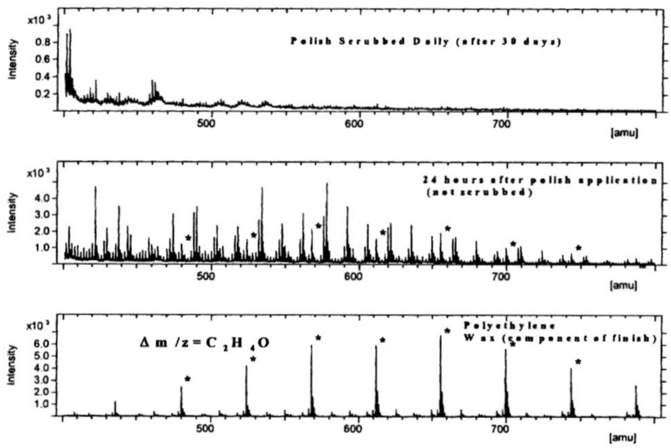

Figure 9 - *Time-of-Flight-Secondary Ion Mass Spectrometry (ToF-SIMS) uses a pulsed, focused primary ion beam to sputter secondary ions from the sample surface. Mass is determined with ToF-SIMS by measuring the flight time of the secondary ions in a drift tube of known length after accelerating the ions to a common kinetic energy. The ToF-SIMS technique is currently being accessed via an external contract laboratory. ToF-SIMS offers several advantages over the in-house quadrupole SIMS instrument. These advantages include mass resolution, mass accuracy, mass range, and imaging capability. Secondary ion images are produced by plotting the intensity of the secondary ion of interest versus XY position on the surface. Dynamic SIMS provides elemental information versus depth.*

Table 1- *Electron Spectroscopy for Chemical Analysis (ESCA)*[1]

	Weight percent												
	Na	F	O	K	CF3	CF2	COOR	COR	CHx	S	P	Si	Zn
Un-scrubbed (fresh)	4.83	24.14	20.18	10.53	5.33	10.66	2.02	5.37	10.15	5.08	1.04	0.23	0.44
Scrubbed Daily after 1x week	4.24	23.82	21.39	10.04	4.01	8.35	2.37	6.08	12.44	4.31	0.84	1.33	0.41
Scrubbed Daily after 1 month	1.35	0.82	32.93	1.56	0.00	0.00	7.84	14.81	35.36	1.20	0.00	3.74	0.38
Polyethylene Wax Emulsion Film	0.00	0.00	28.28	20.05	0.00	0.00	4.46	10.52	31.35	5.24	0.00	0.00	0.00

[1]*Electron Spectroscopy for Chemical Analysis (ESCA) Provides quantitative elemental and chemical state information from the first 20-100Å of a surface. In ESCA, a sample is irradiated with a monochromatic beam of X-Rays. Absorption of X-Ray photons by atoms in the sample causes electrons to be ejected whenever the X-Ray energy exceeds the binding energy of the electron. Photo-emitted electrons leave the sample with excess energy KE, which is measured by the ESCA spectrometer. The respective binding energies can then be calculated from the energy conservation equation:* $Eb = h\nu - KE$. *These binding energies are characteristic of the atomic orbitals, thus permitting ready identification of elements in the sample. Sputter depth profiles provide elemental and chemical state information as a function of depth below the original surface. An energetic ion beam is rastered over the surface to remove material. Profiles are obtained by cycling the ion gun to remove material and collecting ESCA data from the newly exposed surface.*

Because this type of formulation permanently alters the abrasion resistance properties of the polish film surface, it also decreases its utility as a wear layer. Restorers must be burnished after they are applied. High speed burnishing equipment completely removes most restorer formulations.

Dressings are very often simply diluted polishes, perhaps modified with increased wettings agent. Rather than promoting easy removal of the polish surface, to get to the bottom of traffic damage, a dressing brings the bottom of the deep scuff, scratch, or cut up to the film surface by filling in the defect.

Pads

Pad construction and design has followed machine design, and high speed burnishing pad development has reached the stage where generalizations can be made about the relationship between pad/machine combinations and high speed burnish response. Pads vary from very aggressive constructions, usually based on natural fibers, to moderately aggressive constructions, usually based on synthetic fibers. As with spray-buffing, pad selection for high speed burnishing is made simple by using color coding to identify the type of pad. Generally, the darker the pad, the more aggressive with black being the most aggressive and white the least. When selecting pads, it is better to err on the side of being too gentle. A more aggressive pad will emphasize abrasive polish removal (repair), and a less aggressive pad will emphasize abrasive smoothing (gloss build). As indicated by the photomicrograph depicted in Figure 5, the pad selection should be based on machine rotating speed. For example, an aggressive pad should be used with a low speed machine so that the pad can compensate for the lower abrasive polish removal characteristics of the machine. On the other hand, a less aggressive pad should be employed for use with higher speed machines because the abrasive removal of the polish surface by the higher speed machines does not need to be augmented by the pad.

Burnishing Formulations

As in the case of spray-buffing, burnishing requires specially designed formulations for paramount results. The most important features are exceptional durability and slip resistance.

A polish is exposed to incredibly harsh forces from high-speed floor machines. If it lacks film integrity, the floor care maintenance contractor will have to apply new polish frequently. In addition, the user must remember that because high-speed burnishing takes away thinner layers of polish than spray-buffing, it is less effective at removing or "repairing" deep scratches and marks. As a result, a polish employed in a burnishing system must be obviously more durable—more resistant to marking and scratching—in the first place.

A preferred method to making base polishes responsive to high speed burnishing is straightforward: choose a very durable polymer as the finish vehicle and adjust the Polymer/ASR/Wax ratio for good burnish response. This route is confirmed by the experimental evidence presented in Figures 7 and 8 where "durable" coatings based on "harder" polymer vehicles(high glass transition temperature, Tg, and sufficiently crosslinked ionically, covalently or both) can absorb and dissipate the energy of pedestrian traffic and machine maintenance more readily than "less-durable" finishes.

Without a doubt, a softer polish (based on lower Tg polymer and less crosslinking) is easier to abrade, and it will respond very readily to the burnishing process. However, the film formed by such a formulation will be extremely prone to deep scuffing, marking, and soiling. Most important though, a softer polish will not last as long as a polish based on a "harder" more durable polish polymer. The Dynamic Mechanical Analysis data presented in Figure 8 shows that at high temperatures, usually above the glass transition temperature of the polymer comprising the floor polish coating, uncrosslinked polymers or polymers having low crosslink density lose their resilience and structure as stress is applied: energy input to the polish film from foot traffic is not dissipated and breaks down the finish film. Polishes based on durable, highly crosslinked polymer vehicles retain their resiliency so that the energy from pedestrian foot traffic can be readily dissipated. To maintain the best appearance, maintenance personnel will have to burnish constantly. A contractor using a tougher polish can achieve the same level of gloss, and he will not have to burnish for repair as often.

Obviously, the more durable polish offers a savings in labor costs, one of the primary goals of machine maintenance.

Comparative Benefits of Spray-Buffing and Burnishing

Both spray-buffing and burnishing have merits and make them attractive choices in the field of floor maintenance. Spray-buffing does an excellent job of restoring the appearance of the base polish and extending its useful life. The procedure can be done with standard equipment, and it avoids the problems associated with polish buildup and frequent recoating and stripping in high-traffic areas. On the other hand, with burnishing, the high speed of the equipment enables the maintenance personnel to cover much more floor area in a limited amount of time than a spray-buffer will permit. Burnishing also provides ultimate gloss properties, but it does not repair scratches as well as spray-buffing.

The user's decision to spray-buff or burnish depends upon the person's preferences and equipment. If wet-look gloss and rapid floor coverage are desired, the inclination is toward burnishing. Of course, high-speed burnishing units are considerably more expensive than conventional "swing" floor machines, and the would-be burnisher should be aware of this capital expenditure. Furthermore, on floors where quarters are cramped or where there are no long corridors for "straight-line" runs, the unwieldy burnishing units may not perform well enough to justify their expense.

If the floor maintenance person chooses burnishing, the contractor should be aware of two points. For one, high speed burnishers, as have been indicated, remove a very small amount of the polish surface and do not eliminate deeply entrenched scuffs, soil, or marks. Thus, the floor can appear to be glossy but remain dirty. The maintenance contractor must also remember that most floor substrates underneath the polish are to some extent irregular. Since the machine pad traverses over the floor's high spots, high speed burnishing often tends to emphasize these irregularities, which can detract from the floor's appearance.

Floor care professionals have another issue to consider with either type of machine maintenance program. With mop/recoat floor maintenance where a machine is not used, the gloss of the finish acts as an excellent indicator of when recoating is necessary. However, both spray-buffing and burnishing maintain a smooth surface and a constantly high level of gloss that extends the time between stripping/recoating sessions. Consequently, instead of recoating on the basis of gloss, the floor finisher must lay down a new coat of polish at predetermined intervals or after a fixed number of machine passes.

The choice between spray buffing and burnishing is often difficult. Each technique has benefits and drawbacks for different reasons. The ideal situation for floor care people would be a routine that would combine all advantages of both spray-buffing and burnishing. At this time, this is beyond the scope of current technology

Summary

The popularity of machine maintenance in floor care has made formulating floor polish a more complex process. With a mop/recoat maintenance program that does not rely on the use of a machine, the formulator and end user only have to be concerned with the performance properties of the polish. Performance properties are equally critical in finishes for machine maintenance, but they are not the only concern. Formulator and end user alike must make sure that the polish formulation is compatible with the combination of floor machine, pad, spray-buff medium or burnish dressing that will be used.

The demands of machine maintenance require special finishes and systems based on glossy, extremely durable polymer technologies.

Joseph M. Owens[1]

The Effect of Polish Maintenance on Static Coefficient of Friction[2]

Reference: Owens, Joseph M., **"The Effect of Polish Maintenance on Static Coefficient of Friction,"** *Technology of Floor Maintenance and Current Trends, ASTM STP 1448,* W. J. Schalitz, Ed., ASTM International, West Conshohocken, PA, 2004.

Abstract: The static coefficient of friction (μ_s) of polishes, measured by the James machine and ASTM D2047, provides a measure of the inherent slip resistance of the polish film. When in use, polish films are invariably modified by traffic and routine maintenance procedures that could, theoretically, change the inherent slip resistance of the film. Portable tribometers are not appropriate for measuring μ_s of polishes in traffic because they have not yet demonstrated a correlation with the James machine, so they provide no criterion for determining whether the surface is slip resistant.

A protocol was developed using the James machine and a modification of ASTM D2047 to measure μ_s of polish films after pedestrian traffic and routine maintenance had been performed. The maintenance procedures included high speed burnishing, high speed spray buffing, low speed spray buffing, damp mopping, autoscrubbing, deep cleaning (partial stripping), dust mopping, and dry sweeping.

The protocol was extended to test the effect of traffic and maintenance on μ_s. This paper also reports the μ_s changes observed with three compositionally different but commercially typical floor finishes and two widely used commercial polishes of unknown composition.

Keywords: Floor polish, static coefficient of friction, dynamic coefficient of friction, James machine, portable tribometers, slip resistance, damp mopping, mop scrubbing, machine scrubbing, auto-scrubbing, low speed spray buffing, high speed spray buffing, high speed burnishing.

---

[1] President, J. M. Owens, Inc., 128 Oxford Lane, North Wales, PA 19454-4400.

[2] Work conducted at Rohm and Haas Company Research Laboratories, Spring House, PA. Portions of this research were reported at the *International Symposium on Slip Resistance: the Interaction of Man, Footwear, and Walking Surfaces*, held at the National Institute of Standards and Technology, Gaithersburg, MD, October 30-31, 1995, sponsored by ASTM Committee D21 on Polishes.

Background and Introduction

The slip resistance of polishes is determined by measuring their static coefficient of friction (μ_s) by the procedure of ASTM Standard Test Method for Static Coefficient of Friction of Polish-Coated Floor Surfaces as Measured by the James Machine (D2047). This method is used in conjunction with ASTM Standard Practice for Preparation of Substrate Surfaces for Coefficient of Friction Testing, (D4103), and ASTM Standard Test Method for Application of Emulsion Floor Polishes to Substrates for Testing Purposes (D1436). These procedures specify that the polish film is not to be modified in any manner before testing so that the μ_s measured is that inherent to the polish film, rather than some modification of the polish film.

ASTM D2047 provides a conformance criterion of 0.5 μ_s. Polishes that have a μ_s that equals, or is greater than this criterion, when measured in accord with the standard procedure, are deemed to be slip resistant and to provide a standard of safety for polished pedestrian walkway surfaces. The criterion is based on years of μ_s measurements by the protocol incorporated in D2047, correlated with independently accumulated industrial experience of slip and fall incidents and complaints. This experience was gained with polishes that were trafficked and maintained in the normal manner with commonly practiced procedures, even though the μ_s measurements were all made on unmodified polish films.

A logical extension of the above correlation is to posit that a static coefficient of friction of at least 0.5, when measured by any reproducible means, is sufficient for a slip resistant surface. This extension is not logically defensible because the correlation with walkway safety is made with the μ_s of the unmodified polish, rather than the polish film as it is under traffic and as modified by maintenance practices. These static coefficients of friction are not known to be the same. To make the extension logically consistent, it is necessary to determine the μ_s of the polish after it has been modified by traffic and maintenance, since this modified surface is the actual walkway. An unmodified polish surface exists for only a very short time in the functional life of a polish film.

Because the unmodified polish no longer exists as soon as the polish is walked on, or cleaned, or high speed burnished, or spray buffed, the question becomes one of the actual μ_s of the polish film that is being trafficked. This research is intended to answer this question and provide data to evaluate the soundness of the logical extension.

Test Instrument Considerations

The James machine is a large, bulky instrument that is designed for laboratory use, not amenable to being taken to the floor to measure the μ_s of polishes while in use. Though it is possible to make the mechanism of the James machine portable, it is not possible to make portable the 75 pound (34.0 Kg) weights that provide the normal (vertical) force to the leather/polish frictional interface. Because traffic and maintenance is done to floors in place, the measurement of μ_s for polishes in use requires that the measuring device must be brought to the floor, or the floor somehow be brought to the James machine.

Portable tribometers are not appropriate for measuring μ_s of polishes in traffic because they have not yet demonstrated a correlation with the James machine. As such, they provide no criterion for determining whether the surface has retained its slip resistance characteristics. Furthermore, most portable tribometers measure dynamic coefficient of friction (DCOF) or some unknown other undefined property euphemistically called a 'slip index'. (For some portable tribometers, the 'slip index' appears to be DCOF, or some variant of DCOF. There is no established criterion for slip resistance based on 'slip index', other than the mythic - and perhaps erroneous - impression that higher values should be better.) It has been established [1] that DCOF has no bearing on whether a slip will happen, though DCOF has pertinence to whether a slip will progress into a fall. Accordingly, there is currently no objective criterion (and probably no hope for one) that establishes a minimum value of DCOF for a safe walkway surface. Comparison of DCOF measurements before and after floor maintenance cannot tell us whether the maintenance operation has changed the slip resistance of the polish film, since DCOF is not pertinent to whether the polish is slip resistant. That is, traffic and conventional maintenance procedures may change the DCOF of a film without changing μ_s or slip resistance, or they may change μ_s without changing the DCOF. Only the static coefficient is pertinent to the slip resistance of a polish.

The few portable tribometers that are capable of measuring μ_s, such as the Brungraber Mark I tester and the Liberty Mutual Horizontal Pull slip meter, have shown a poor correlation with μ_s data from the James machine, even when they use the same polish and specification leather for the polish/sensor frictional interface. This poor correlation is thought to be due to the significantly smaller normal force (vertical weights) applied to the frictional interface by these portable machines and the lack of compression of the leather sensor. For whatever reason, this lack of correlation means that there is no way to relate the data from these portable instruments with the experiential data on the slip resistance of polishes. The Mark I machine (preferred because it is easier to use than the Horizontal Pull machine) would be able to tell if μ_s increased or decreased as a result of traffic or polish maintenance but would not be able to indicate if the polishes were still slip resistant.

Coefficient of friction data generated by ASTM D2047 is the only appropriate means for determining if the ravages of pedestrian traffic and conventional maintenance significantly alter the slip resistance of a polish film and, perhaps, render it less than safe for pedestrian traffic. For this study the lack of portability for the James machine can be addressed by designing test protocols that bring the trafficked or maintained polish and flooring material to the machine.

Trafficking and conventional floor maintenance are large-area processes. Modifying these processes so that they can be done on a small area in the laboratory would be an ideal solution that would make the product tiles available to the James machine. Reducing these processes to a laboratory scale so that their effects could be more easily (and cheaply) measured and manipulated has defied scientists for at least the past half century. Part of this difficulty is because the only way we can know we have successfully miniaturized these large scale processes and their effect on μ_s is to compare the lab results with actual floor results. We have μ_s data from neither the full scale processes nor lab scale approximations of these maintanance processes. The work of this paper was targeted on obtaining μ_s data on polish films as a function of large scale pedestrian traffic and maintenance.

Floor Test Area

Guidance in the setup of a test area for traffic and maintenance of floor polishes is provided by ASTM Standard Practice for Rating Water-Emulsion Floor Polishes (D 3052), and ASTM Standard Practice for Evaluation of Spray-Buff Products on Test Floors (D3758). (At the time this study was initiated, the 1985 edition of standard ASTM D3758 was undergoing revision. The proposed revision, approved in 1995, was used as the basis for the protocol of this study.)

An ideal traffic test area for polish evaluation provides a straight passage for pedestrians, with no turns or forced changes in direction or gait. It will have no opportunities for traffic diversion, so all of the pedestrians will sequentially pass in both directions over each of the polishes being tested. The area will be long enough to allow testing of multiple finishes, including control polishes with known performance in the field. Each polish should be allocated at least 48 inches (1.22 m) along the directions of traffic flow. Traffic density will be sufficiently high to provide aggressive wear to the finishes, though not unusually abrasive or dirty, mimicking the traffic and soils the test finishes are intended to encounter. Traffic density will be low enough to allow timely monitoring of changes in polish performance and appearance but high enough for rapid generation of results. The flooring material will be compatible with the finishes and chemicals used in their maintenance and evaluation.

Most of these conditions were met by an internal corridor connecting two buildings at the Rohm and Haas research facilities in Spring House, Pennsylvania. The exception to our ideal was that the flooring was not amenable to μ_s determinations by the James machine. Also, one end of the corridor ends in a "T" intersection with a cross-corridor, and the other end is a pair of fire doors (automatic closers to maintain building fire integrity) that open inward onto the corridor. Thus both ends involve directional or gait changes. The corridor is 6.83 feet (2.08 m) wide and 48 feet (14.63 m) long and the traffic density, presumed to be high, was to be determined. Because neither of the buildings connected through this corridor seemed to fill up with pedestrians during the workday, we presumed that traffic was equal in both directions.

An alternate traffic route was available for those times when this corridor would be denied to pedestrians by our work. Since the alternate route is longer, our closing this passage was not popular with our pedestrians; so much of the work was done at night or weekends when the complaints were not as numerous. We also provided long periods when no polishes or procedures were being tested so that the pedestrian traffic would not habitually divert to the alternate route.

The floor tiles in the corridor were made to be removable so that they could be taken to the James machine. The installed tiles were removed, and the residual adhesive scraped from the concrete "slab on grade" floor. These were replaced with 12" x 12" x 1/8" (20.5 cm x 30.5 cm x 0.32 cm) black and white official vinyl composition tiles[3] (OVCT), which were press-fit on the floor without adhesive. The rows of tiles directly abutting

[3] Available from Consumer Specialties Products Association, 900 17th St., NW, Washington, DC, 20006 (202) 872-8110.

both walls of the corridor were trimmed to fit under the cove molding and still provide lateral tension to hold the remainder of the tiles in place without moving.

Traffic Count

The tiles were stripped of their factory finish in the conventional manner[4]. After air-drying, three coats of polish formulation N-32-2 (formulation details in the Appendix) were applied with a string mop. The polish was allowed to dry overnight before being subjected to traffic. No maintenance, other than daily dust-mop sweeping, was done on the polished floor. Polish performance was evaluated twice weekly with visual assessments of gloss, scuff marking, and 60-degree gloss meter readings taken on each tile in the floor.

The corridor was fitted with a custom photo-electric cell and counter to determine the average daily traffic density. It was necessary to discard the data from the first six weeks of counting pedestrian passes, because the lure of playing with the light beam and counter (it clicked with each count) could not be ignored by our professionally curious pedestrians. After 12 additional weeks of traffic, the traffic count averaged 850 pedestrian passes per workday, with a range from 750 to 1000. Over weekends, there were an additional 225 passes (maintenance personnel and assorted workaholics) for a total traffic load of about 4500 pedestrian passes per week. This level of traffic density is thought to be moderate or moderate-to-heavy. Higher traffic density corridors were available on site, but they did not have the advantage of all traffic passing over the entire length of the test area (offices, labs, or intersecting corridors diverting some of the traffic).

Traffic Profile and Wear Test Map

The row of tiles along each wall of the corridor received almost no traffic, as denoted by gloss retention and the absence of scuff marks on these polished tiles. Traffic was centered and concentrated in a five foot (1.52 m) width of the corridor, with the greatest traffic over the central three feet (0.91 m) of corridor width. The cross-sectional traffic density appears to be a slightly skewed bell-shaped curve with a maximum in the center of the corridor width. The two rows of tiles adjacent to the central row are estimated to experience about 20 to 25 percent less wear than the central row, and the outer two rows (second row to the right and left of the central row) about 20 to 30 percent less wear than the more central adjacent rows. A traffic density cross-section profile (center of the corridor length) is shown in Fig. 1. A traffic pattern map is shown in Fig. 2.

The end of the corridor that intersects with the cross-corridor shows a flaring of traffic toward both walls, though skewed to the North side. This is in agreement with visual observations of pedestrians who must turn either right or left at the end of the test area, and more of the pedestrians turn to their left or enter the test corridor from the North.

---

[4] 3M black High Pro stripping pad under a 17 inch (43 cm) diameter, 175 rpm swing floor machine, with S.C.Johnson's *Step Off* used at 1:4 dilution. After double stripping, the floor was rinsed with clear water, vacuumed dry, and then damp mop rinsed twice with clear water.

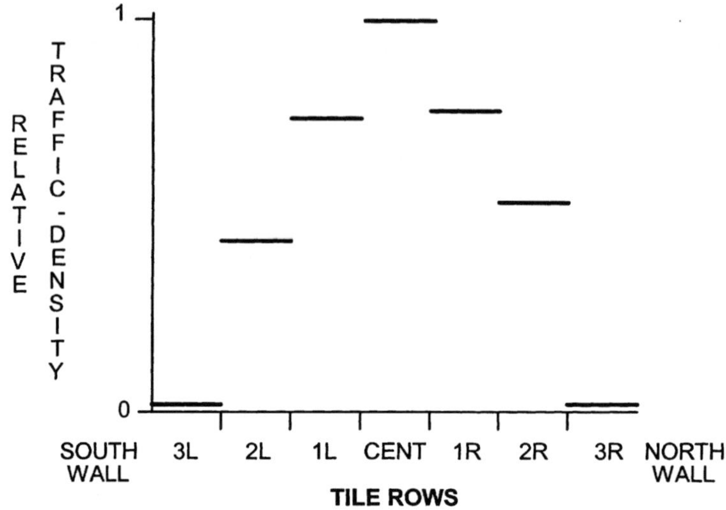

Fig. 1–*Traffic density profile across the Corridor.*

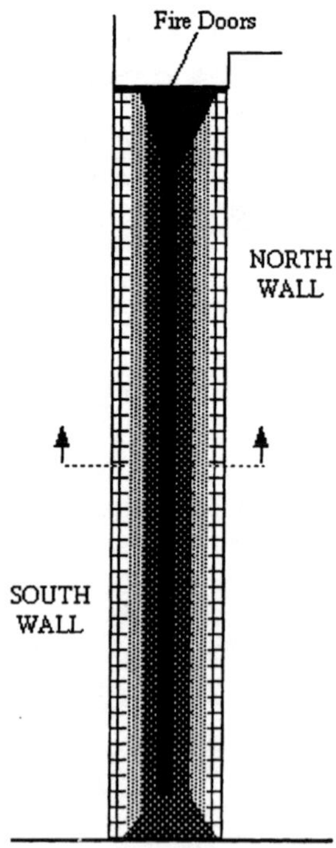

Fig. 2–*Map of test area traffic pattern.*

The turning of traffic, or fanning of traffic density, begins about three feet from the end of the test corridor. Because this end of the test corridor receives a turning or torquing traffic different from the body of the corridor, the last four feet (1.22 m) of the test corridor before the intersection were not used in this testing.

At the other end of the test corridor, the traffic pattern, as denoted by scuff marks and loss of gloss, also changed as pedestrians came to a full stop and then rocked back or shuffled their feet to pull open the fire doors. This change in traffic pattern occurred about five feet (1.52 m) from the end of the corridor, with maximum traffic damage about three feet (0.91 m) from the doors, so the final eight feet (2.44 m) of this end of the test area were excluded from use for this testing.

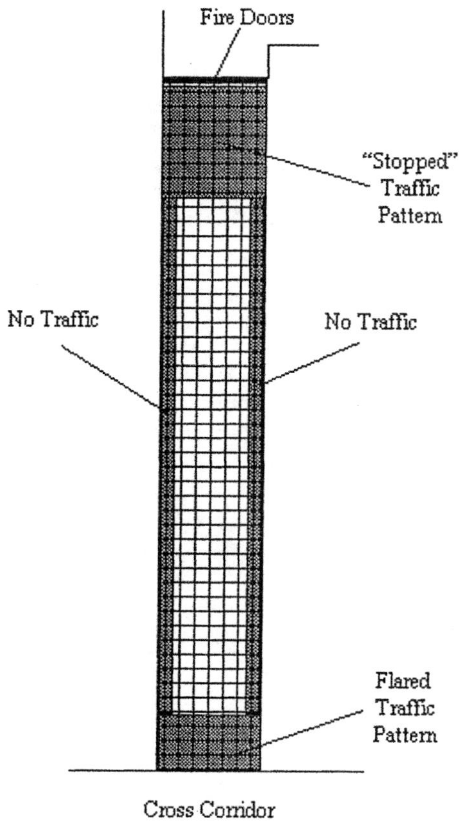

Fig. 3–*Test area map.*

After subtractions from the corridor for the ends where the traffic pattern changed and the tile rows adjacent to the walls devoid of traffic, the effective test area is five feet (1.52 m) wide and 36 feet (10.97 m) long, or 180 square feet (16.72 m^2). A map of the test area is shown in Fig. 3.

Pedestrian traffic for these tests was supplied by our unwitting co-workers. They were not representative of the general population[5]. Our population consists of about 70 percent males, with ages distributed from mid-20's through late-50's and a median age of 39. The average IQ on site was near 120, which was not expected to skew the type of wear that is imparted to a floor finish, but running, and even very fast walking, is very rare for this group of professional scientists. The weight to height ratio of this population is lower than the general population which means that individual foot strikes should be lighter than the general population.

The footwear of our pedestrians is also atypical, since about 90 percent wear company-supplied (free) safety shoes. Open footware and shoes with cloth uppers are forbidden on-site as a safety measure. Most of the shoes are relatively new, since personnel may select a new pair at least once each year. The heels and soles of these safety shoes have a broad range of hardness but are not as soft as sneakers or tennis shoes. The sole and heel rubber is uniformly oil and solvent resistant. The harder heel and sole materials tend to increase the impact energy transferred to the polish with each individual foot strike.

The Effect of Polish Application on SCOF

In North America all large scale polish applications are done with a string mop.

Though ASTM D2047 stipulates that polishes be applied to the test surfaces by the method of ASTM D4103, which in turn specifies using one of the methods of ASTM D 1436. These are all hand applications, or mechanized variations, and are laboratory procedures; they are designed to ensure only that an even, coherent film of polish is applied, so that the μ_s testing will have the polish film surface as the variable in the test frictional interface. The laboratory application procedures attempt to avoid, rather than replicate, the random variations that are presumed to be inherent to string mop application of polishes. These variations include missed spots (called "holidays"), random fluctuations in film thickness, and wetting and leveling ridges. We took, as our first objective, to test whether string mop application caused any changes in the μ_s of a polish film, relative to the laboratory applications.

Polish Selection

The polish formulations chosen for this test were N-32-2 and N-32-25. They are examples of the most widely used technology in the world wide floor polish industry, based on a conventional metal crosslinked, styrenated-acrylic copolymer of modest hydrophilicity. The formulations contain conventional synthetic "waxes" with a polymer/alkali soluble resin (ASR)/wax ratio of 75:10:15, a composition typical of minimal maintenance polishes. Formulation details are presented in the Appendix.

At 15 percent total solids, formulation N-32-2 represents the low end of the solids spectrum for industrial and institutional polishes. Low polish solids should emphasize any deviations in μ_s caused by incomplete coating of the floor. Though multi-coat applications of low solids polishes are standard in the industry and they are routinely

---

[5] Our pedestrian population data was provided by Rohm and Haas Human Resources department for the Spring House site and assumed to be applicable to the occupants of the connected laboratory buildings. General population data is from the 1990 United States census [2].

recommended in label directions, we will test the effect of single and multi-coat applications on the μ_s of polishes.

Formulation N-32-25 is the same formulation as N-32-2 at 25 percent total solids. This represents the high end of the solids content range for the majority of industrial and institutional polishes. A high solids level is anticipated to emphasize deviations in μ_s caused by inadequate leveling, and wetting, or uneven polish application. At the time this work was initiated, the median polish solids in the industrial and institutional market was 18 percent.

These particular polishes were selected primarily because they are representative of the technology used in the majority of industrial and institutional floor polishes. Their selection also allowed us to use the large data base of μ_s measurements by ASTM D2047 because these polish formulations are used almost every day as a check of the calibration of the James machines at Rohm and Haas Company's Floor Care Research labs. For formulation N-32-25, this data base consists of 24 620 determinations of μ_s, and for N-32-2, 6 710 determinations of μ_s. All of these determinations were made on polishes applied in accord with ASTM D4103. Almost all (about 95 percent) of the above determinations of μ_s were made with the same James machine used in this testing. In this report the data base average μ_s values are reported as "archive μ_s."

Procedure

The test corridor was stripped of polish, rinsed, and allowed to dry, as described earlier[4]. The floor was divided into two areas (marked on the baseboard at the midpoint of the corridor length), so that each area was 24 feet (7.32m) long by 7 feet (2.13m) wide. Because the polish was not going to be exposed to traffic before μ_s testing, the ends of the corridor, which were excluded from traffic tests because of changes in the traffic pattern, were included in this test.

Four bare tiles from random locations in each of the two areas were pried up and moved to the constant temperature/humidity room (72° F, 55% RH) for μ_s determination in the procedure of ASTM D2047. After measurement, the eight tiles were press-fit back into the floor.

One coat of formulation N-32-2 was applied over the entire corridor and allowed to air dry for about 30 minutes (the fire doors were blocked open[6] to aid in air circulation). Polish application was done with a clean 12 oz cotton string mop in a casual manner. That is, no particular efforts were made to apply an exceptionally thick or thin coating. Previous experiments have demonstrated that this method of application results in a coverage rate slightly less than 2000 square feet per gallon. This application rate is believed to be typical and is the rate commonly recommended by commercial industrial polish label directions.

When the first coat was dry to touch (about 20 minutes), a second coat of the same formulation was applied to the east half of the corridor length. This coat of polish was applied in the same manner as the first.

---

[6] Local code requires that when the fire doors are blocked open they must be guarded so that they can be closed immediately if an emergency should arise. The guard also warned the odd obstinate or preoccupied pedestrian away from the wet floor.

After the second coat of polish had been allowed to dry for two hours, eight tiles were removed from random locations in each half of the test area. The tiles were marked on the back to identify which end of the test area they came from, though this was easily distinguished by the higher gloss of the double-coated tiles. These tiles were transferred to the constant temperature/humidity room for μ_s testing.

After measuring the μ_s by the procedure of ASTM D2047, the 16 tiles were press-fit back into the test floor.

The test floor was again stripped of polish, rinsed, and allowed to dry[4], and the above procedure was repeated in its entirety, with formulation N-32-25 used in place of formulation N-32-2.

Results and Conclusions

The results of the μ_s determinations are presented in Table 1. With the exception of the data marked "archive μ_s," each data point is the average of μ_s determinations on eight tiles.

Table 1–μ_s *of mop-applied polish*[7].

Uncoated tile	0.39 +/- 0.02	0.37 +/- 0.02
Polish	N-32-2	N-32-25
Archive μ_s	0.55 +/- 0.01	0.55 +/- 0.01
String Mop Application		
One Coat	0.48 +/- 0.04	0.55 +/- 0.02
Two Coats	0.55 +/- 0.02	0.54 +/- 0.02

The lower μ_s value of the bare tiles in the test of formulation N-32-25 is statistically insignificant (the difference may not be real). It is, however, in the direction anticipated. Previous experience with μ_s determinations for uncoated flooring materials has shown that unpolished vinyl composition tiles have μ_s values from 0.32 to 0.42, depending on the age and previous use of the tile, as well as variations in manufacture. More porous tiles, such as those stripped more vigorously or more often, or tiles that have been subjected to heavy, gouging traffic, are typically at the lower end of this range. Though these were new tiles, they had been subjected to heavy stripping to remove the factory finish, and the tiles in the N-32-25 application test were stripped one time more than the same tiles in the N-32-2 application test.

Multi-coat string mop polish application yields the same static coefficient of friction as polishes hand-applied per ASTM D4103 and ASTM D1436.

---

[7] The error limits are one standard deviation of the mean.

The lower μ_s of the single coat application of low solids polish (N-32-2) is indicative of an incomplete film over the somewhat porous tile. With some of the very low μ_s values bare tile is apparently still exposed to become part of the frictional interface. Adherence to the common practice of applying at least two coats of polish, particularly if it is a low solids polish, should avoid this reduction in μ_s, though a single coat of this low solids polish is still slip resistant.

Though it is only indirectly related, the gloss data that was also collected during this test is informative. Photovolt 60° reflectance gloss data is shown in Table 2. The procedure of ASTM Standard Test Method for 60° Specular Gloss of Emulsion Floor Polishes (D1455) was used, with the exception that the string mop application procedure of this testing was the application method. (The gloss data in Table 2 for "archive" gloss is from the large database on these formulations and was not generated in the testing reported here.)

Table 2–*Gloss of mop-applied polish*[7].

Uncoated tile	8 +/- 4	6 +/- 4
Polish	<u>N-32-2</u>	<u>N-32-25</u>
Archive	70 +/- 3	81 +/- 3
String Mop Application		
One Coat	28 +/- 5	56 +/- 4
Two Coats	60 +/- 4	78 +/- 4

The gloss values are as expected, showing higher gloss in those instances where more polish was applied to the substrate. The higher gloss of the tiles coated by the laboratory method of ASTM D1436 (archive data) indicates that it provides a heavier film of polish than does a two coat string mop application.

If we juxtapose the data from Table 1 with that in Table 2, the data shows that there is *no* correlation between gloss and μ_s. If consideration of the uncoated tiles is included, it may appear that μ_s increases as a function of gloss, but that is bogus.

Since the conventional mode of string mop polish application preserves the μ_s of the polish film, relative to bench applications, we can now turn to determining if modification of the polish film by simple maintenance procedures has any impact on μ_s.

The Effect of Damp Mopping on SCOF

Besides simple sweeping, the oldest, and simplest polish maintenance procedure is damp mopping. This is typically done with clear water, lightly applied with a string mop and then taken up or simply allowed to dry. A bucket with wringer is used to control the transfer of liquid to and from the floor.

Damp mopping is used to remove water soluble soils, or fine, loose dirt from the floor, and it is not appropriate for heavy soils. Because it is a labor intensive (and therefore expensive) procedure, it is now usually restricted to small areas of the floor, or "spot cleaning," rather than large expanses.

Because the floor polish is made wet by damp mopping, μ_s and polish slip resistance is very low while the procedure is being done. (Rather than the polish film being the variable in the frictional interface, a wet floor has a mobile layer of water as the variable in the frictional interface.) It is standard janitorial practice to place warning signs, safety cones, or other pedestrian impediments to keep traffic off a wet floor. Every wet floor should be considered unsafe for pedestrian traffic.

We are not concerned here with the status of μ_s while the floor is wet from damp mopping. The James machine is known to give erroneous data for wet surfaces. (When wet, the specification leather of ASTM D2047 appears to act as a suction cup so that μ_s values are exceptionally high. They are often higher than the μ_s values obtained from the same polish when dry, which is contrary to the general pedestrian experience with wet floors.) We are concerned about the effect of the damp mopping maintenance operation on the slip resistance of the floor when it is again opened to pedestrian traffic.

It may be hypothesized (based on no data in support or to the contrary) that damp mopping could change μ_s by extracting chemical constituents from the polish film that are either water soluble or sufficiently hydrophilic. Examples are the surfactants and residual coalescing solvents in polish formulations. This change in film composition could change the static coefficient of friction. Even if materials were not extracted from the film, the chemical characteristics and μ_s could be changed by water penetration into the polish film surface, causing the film to swell and expose chemical constituents that had been buried in the film.

Polish Selection

As a relatively hydrophobic control polish, we again used N-32-25 in this test.

A more hydrophilic polish technology is represented by formulation DP-2-4. It is based on the second most widely used technology in the world wide floor polish industry. The formulation has a polymer/ASR/wax ratio of 85:5:10, optimized for both minimum maintenance and spray buffing maintenance (see below) procedures. Though it provides higher crosslink density and greater film hardness and toughness than N-32-25, DP-2-4 is based on a more hydrophilic styrenated-acrylic copolymer. It should have a greater tendency to exhibit the hypothesized changes in μ_s that may be a consequence of damp mopping. The formulation details for DP-2-4 are given in the Appendix.

Because formulation DP-2-4 is used extensively as a control polish for both internal and external wear tests, there is a moderately extensive body of data available on it from 626 determinations of μ_s per ASTM D2047.

A third polish was included in this part of the testing. This polish was claimed (at the time this testing was initiated) to be the largest selling polish formulation in the world. Because it is a proprietary, commercial product with a composition controlled by the manufacturer, formulation details are not available. It will simply be designated as polish "Commercial C." "Commercial C" has also been used extensively as a control polish for internal and off-site wear testing, so there is a large body of data on μ_s determinations. This archived data indicates that the formulation has changed at least twice in the time that we had been measuring its μ_s. Of the 1452μ_s measurements, 505 were on the variation of the formulation that was included in this testing. The slightly broader standard deviation of the mean for the archive data on this variant of "Commercial C" may be an indication that there have been minor formulation adjustments (and corresponding minor changes in μ_s) of which we were not aware. "Commercial C" is formulated at 17 percent solids.

Procedure

The wear test corridor was stripped of polish as described earlier[4], and the test area was divided into three sequential 12 foot (3.66 m) lengths.

After the tiles had dried, 28 tiles were removed from the center of each of the three test areas. Each set of 28 tiles was coated with the respective polishes described above, by the procedure of ASTM D4103 and ASTM D1436. The slower dip coat method of ASTM D 1436 (Method A) was used instead of the hand applicator method (Method B) because it is known (based on initial gloss) to apply a heavier coat of polish. Heavier coats of polish would be expected to accentuate the hypothesized mechanisms for changing μ_s, if they occur. Two coats of polish were applied and allowed to dry overnight.

Four tiles from each set were set aside in the constant temperature/humidity room for determining the baseline μ_s of the polishes, to be compared with the archival values. The spaces for these tiles were filled with other miscellaneous OVCT tiles at hand. The remaining tiles were replaced in the test floor.

The entire test floor was damp mopped (16 oz cotton string mop) with clear water. The water was changed and a different clean string mop was used for each polish.

The floor was allowed to dry for 20 minutes (fire doors blocked open[6] to enhance air circulation), and the damp mopping with clear water was repeated. The same mop associated with each polish in the first damp mopping was used again.

After a third repetition of the damp mopping the initial set of 24 tiles of each polish was removed to the constant temperature/humidity room for μ_s determination by ASTM D 2047.

Results and Conclusions

The μ_s data is presented in Table 3. The archive baseline data is the average from the previous μ_s history for each polish. The baseline data is the average μ_s from the four tiles of each polish that were not subjected to damp mop maintenance. The damp mop μ_s data is the average from the 24 tiles of each polish that were subjected to damp mop maintenance.

The damp mopping procedure leaves the μ_s of polishes essentially unchanged.

The apparent increase of μ_s for "Commercial C," though statistically insignificant is suggestive, so it was pursued.

The extraction hypothesis was further tested in the laboratory by preparing four tiles for μ_s determination per ASTM D 2047, measuring μ_s, and then placing the tiles in shallow pans of distilled water to soak for one hour. After air drying overnight, the value of μ_s for "Commercial C" was found to increase (see Table 3). The other polishes, which had been unaffected by damp mopping, were also insensitive to the laboratory water soak. The statistical significance of the change in μ_s before and after water soaking is weak, but it gives some credence to the idea that extraction of residual hydrophilic materials from "Commercial C" films by wet cleaning procedures causes μ_s to increase.

The formulation details for "Commercial C" are not known, so it is not possible to identify what hydrophilic or water soluble material or combination of materials in the formulation may be responsible for this change in μ_s, assuming the change is real and not just a random artifact.

Table 3–μ_s *of clear-water damp mopped polishes*[7].

Polish	N-32-25	DP-2-4	Commercial C
Baseline μ_s			
Archive	0.55 +/- 0.01	0.59 +/- 0.03	0.56 +/- 0.04
Determined	0.55 +/- 0.01	0.59 +/- 0.03	0.56 +/- 0.03
After Damp Mopping	0.55 +/- 0.02	0.59 +/- 0.04	0.58 +/- 0.03
Before Laboratory Water Soak	0.53 +/- 0.02	0.57 +/- 0.03	0.56 +/- 0.03
After Laboratory Water Soak	0.53 +/- 0.03	0.58 +/- 0.03	0.60 +/-0.02

The Effect of Mop Scrubbing with Detergents on μ_s

Clear water is surprisingly effective as a cleaning agent, and the string mop in a damp mopping procedure is a highly versatile instrument. However, when faced with oily or more aggressive soils than the water soluble or light, loose soils for which damp mopping is best suited, custodians escalate the battle by using more aggressive cleaning procedures. The escalation takes two forms. The first is the use of surfactant solutions and the second is the use of more aggressive mechanical agitation of the cleaning solution on the floor.

While pure water is capable of dissolving water soluble stains and soils, and dispersing small particle size hydrophilic soils (such as dust and loose dirt), it is actively repelled by oils, greases, and other hydrophobic soils.

Surfactants in cleaning solutions assist in removing hydrophobic soils by coating such soils and rendering the surfaces hydrophilic. This not only negates the inherent repulsion of water from the formerly hydrophobic surface but also makes the soil particle easily dispersed in the water. The stability of the dispersion is a function of the size of the surfactant-coated particle. (Because only the surface of the soil particle is rendered hydrophilic, solution in the water phase is not possible.)

Industrial and institutional polish films are intentionally mildly hydrophobic. This characteristic is necessary to give the films resistance to the water and detergent solutions that they will encounter in their working life. Hydrophobicity is imparted to the film by the hydrophobic monomers used in manufacture of the polymer, and by formulating the polish with hydrophobic plasticizing solvents and synthetic polymers which function as waxes. Care is taken to preserve some hydrophilicity in the film because this property relates to the durability (soil resistance and heel mark resistance, in particular) of the polish. It is also necessary that the film be sufficiently hydrophilic that water, or whatever aqueous cleaning solution is used, will wet out the floor. The cleaning solution must be able to surround the soil particles, including getting between the film surface and the soil, for it to remove water insoluble soils. This would be impossible if the film were so hydrophobic that water beaded up.

Because surfactant solutions are designed to coat and disperse hydrophobic soils, they may also be capable of altering the somewhat hydrophobic polish film in such a way that slip resistance is changed by the cleaning operation. In this part of the study we will test to see if the use of surfactants causes a change in the μ_s of floor polishes.

Polish and Cleaner Selection

We make no claims to having mastered the art of formulating floor cleaners, so a proprietary commercial floor cleaner, "Commercial F," was selected for this study. The formulation is a proprietary blend of non-ionic and anionic surfactants, with alkali and alkaline builders (to keep the pH high, enhance wetting, and ensure the solubility of anionic surfactants).

"Commercial F" is a concentrate formulation to be diluted by the end user in accord with the condition of the floor to be cleaned. We used "Commercial F" at two different concentrations, both per the label directions. The 1:40 dilution is for general cleaning of moderately soiled floors. A higher concentration (1:20 dilution) is recommended for particularly heavily soiled floors or where germicidal properties are needed. The more dilute form is thought to be more typical of general use, but floor finish films will sometimes encounter the more concentrated solution, so both were included in this testing. The dilutions were made by volume, rather than weight, because it is presumed that this would be typical of the way custodians would handle this product.

Not knowing the identity or proportions of the chemicals in "Commercial F," we cannot make judgments about the causes if its use were to change μ_s. After discussions with internal experts, a study of the patent literature, and some chemical analysis of commercial products, we devised a cleaner formulation that is representative of the most common floor cleaners.

This is based on a blend of non-ionic surfactants because surfactant blends are reported [3] to be more efficient at cleaning the wide range of hydrophobic soils encountered from real traffic. The solids of the formulation were set at 0.5 percent, because a survey of commercial cleaner formulations showed this to be the upper end of the range of recommended "use dilutions." The formulation details, and a description of the ingredients, for cleaner formulation NIO-0.5 are given in the Appendix.

Since damp mopping with clear water had no effect on μ_s of the polishes used in that study, the same polished tiles were again used in this study.

Procedure

The tiles used in the test of the effect of clear water damp mopping on μ_s were returned to the test floor for use in this testing.

Each of the cleaner solutions was charged to a four gallon bucket equipped with a squeeze wringer, and a new, clean 16 oz string mop was assigned to each solution.

The NIO-0.5 solution was applied over the entire test floor at a rate of 300 to 500 square feet per gallon. The solution was then scrubbed over the floor with the string mop and picked up with the wrung-out string mop. The floor was rinsed once with clean water (same mop as used in applying the cleaner solution, after being rinsed), with an application rate of 300 to 500 square feet per gallon, and this rinse was picked up with the wrung-out string mop. The spent rinse was discarded, the bucket and mop rinsed out, and the bucket again charged with clear water. The floor was damp mopped with this clean water (again, the same mop), and allowed to air dry for one hour. The cleaning, rinsing, and damp mop rinse steps were then repeated. After the second air drying period, the 28 coated tiles of each polish were taken up from the floor and transferred to the constant temperature/humidity room. The polish film μ_s was determined the next day by the procedure of ASTM D 2047.

After μ_s determinations for the NIO-0.5 cleaner effect were completed, the tiles were returned to the test floor and the testing was repeated with "Commercial F-1/40."

After the μ_s determinations for "Commercial F-1/40" were completed, the tiles were returned to the test floor and the testing was repeated with "Commercial F-1/20."

Results and Discussion

The data from the study of the effect of mop scrubbing with detergents on polish μ_s is presented in Table 4. Because the same polish film surfaces were used for each test, the results show the *cumulative* effect of the tested procedure and chemicals along with the effect of all of the chemicals that had preceded it.

The data is the average of the μ_s determinations on all of the tiles of each polish, rather than treating the 28 tiles of each polish as seven sets of four. This treatment of the data has no effect on the average value of μ_s, but it narrows slightly the standard deviation of the mean and the confidence interval calculations.

Mop scrubbing has no effect on the slip resistance of the polish films.

We previously suggested that damp mop cleaning with clean water could remove some (unknown) hydrophilic materials from the film surface of "Commercial C." There was a statistical suggestion that this hypothesized extraction increased the value of μ_s. The wet mop scrubbing procedure, which puts more water in the film, and for longer times, would be expected to continue this process and extract even more of the hydrophilic materials, further increasing μ_s. This expectation was not reflected in the data so either the expectation was based on an erroneous hypothesis, or the presence of surfactants in the aqueous cleaning solution caused the removal of other materials from the film or film surface, which compensated for the loss of hydrophilic materials into the water.

Table 4—μ_s *of polishes after mop scrubbing*[7].

Polish	N-32-25	DP-2-4	Commercial C
Baseline μ_s			
Archive	0.55 +/- 0.01	0.59 +/- 0.03	0.56 +/- 0.04
Determined[a]	0.55 +/- 0.02	0.59 +/- 0.04	0.58 +/- 0.03
After Mop-On Cleaning[b]			
NIO-0.5	0.55 +/- 0.02	0.59 +/- 0.03	0.58 +/- 0.03
Commercial F-1/40	0.55 +/- 0.03	0.59 +/- 0.03	0.57 +/- 0.02
Commercial F-1/20	0.55 +/- 0.02	0.59 +/- 0.03	0.58 +/- 0.02

[a] From Table 3, after damp mop cleaning; data is the average μ_S determination on 24 tiles.
[b] Data is the average μ_S determinations from 28 tiles.

The extraction soak experiment described previously was repeated with "Commercial C" and cleaner solution "Commercial F-1/20" with no change in the value of μ_s. The net result is that the presence of surfactants in the mop-cleaning solutions, which have the potential to remove unbound hydrophobic materials from the film, did not effect the slip resistance of the polishes.

Like damp mop cleaning, mop scrubbing is highly labor intensive, and so both are expensive operations that are largely restricted to spot cleaning and are rarely used to clean the entire floor of a facility. Though mop scrubbing is much more efficient at removing soils than is damp mopping, it is still relatively inefficient at cleaning floors that have been heavily soiled with oil or grease. The mechanical action to break up these soils must be provided by the worker on the dry end of the mop, which is not only tiresome but also relatively inefficient. Much greater mechanical efficiency is obtained when the agitation is provided by machinery. Every janitor has access to a 175 rpm swing floor machine that is specifically designed for this very purpose, so we will now consider the effect of mechanical scrubbing procedures on the slip resistance of polish films.

The Effect of Mechanical Scrubbing on μ_s

Mechanical scrubbing provides aggressive scrubbing agitation of the cleaning solution to break up oily and greasy soils mechanically so that they can be dispersed.

Mechanical scrubbing, as commonly practiced in the industry, is based on the use of a heavy, rotary 175 rpm floor scrubbing machine, commonly called a "swing" machine, with a moderately aggressive brush or scrubbing pad in full contact with the floor. The full weight of the machine rests on the turning pad. The characteristic swing motion of the machine is provided by the operator skillfully tilting the machine so that a disproportionate amount of weight is applied to the leading or trailing edges of the rotating pad or brush. This drives the machine to the right or left. Tilting the machine to the right or left will drive the machine forward (away from the operator) or back. The technique is readily learned, but proficiency requires a couple of hours of practice.

Mechanical scrubbing requires the same equipment as does the mop scrubbing procedure, with the addition of a floor machine and its moderately aggressive brush or scrubbing pad. It also requires a slightly longer time than does mop scrubbing, but it is much more efficient at removing heavy or oily soils. This improved cleaning efficiency may increase the probability that the process will remove polish components from the film that will change its μ_s.

Polish and Cleaner Selection

The same polish formulations that were used in the damp mopping and mop scrubbing evaluations of μ_s are appropriate for this study. The same cleaner formulations used in that study are also appropriate for this, because the maintenance process variables being evaluated are those of even longer chemical residence times and more vigorous mechanical agitation.

Procedure

The entire corridor was stripped of polish as has been described earlier[4]. The wear test area was divided into three equal 12 foot (3.66 m) lengths, and 28 tiles in each area were randomly selected and marked on a corner with a pencil. This mark on the bare tile was visible through the clear polish coats that were subsequently applied.

Three coats of the designated polishes were mop-applied to each area, with 45 minutes drying time allowed between coats. After the last coat was applied, the floor was left to dry overnight. The polishes applied to the ends of the wear test area (N-32-25 near the fire doors, DP-2-4 near the cross-corridor) were also applied to the adjacent ends of the corridor with different traffic patterns (see Fig. 3). Each polish application was also extended to the rows of tiles abutting the corridor walls so that the entire corridor was polished.

The next day, four of the marked tiles of each polish were removed for μ_s determinations to provide baseline data. The empty spaces were filled with (marked) miscellaneous tiles.

Cleaner solution NIO-0.5 was applied over the entire corridor, at an application rate of 300 to 500 square feet per gallon. The floor was scrubbed with a Hako 175 rpm 17 inch (43 cm) diameter floor machine, affixed with a pad driver and 3M red scrubbing pad[8].

The machine was run in the conventional manner so that each row of tiles (progressing down the corridor from the fire doors) was passed over by the head of the pad moving in one direction, then the body of the pad while swinging in the opposite direction, and finally by the heel of the pad while swinging back in the original direction. Each tile received a minimum of three scrubbing passes, with some overlapping of passes.

After the entire corridor had been scrubbed, the spent, foamy cleaning solution was removed with a wrung-out string mop. A rinse of clean water was applied at a spread rate of about 300 square feet per gallon. The rinse was taken up with a wrung-out string mop,

---

[8] Available from 3M Corporation, Minneapolis, MN, but sold only through janitorial distributors.

and the floor was again rinsed with clean water in a damp mopping operation. After allowing three hours to dry, the above procedure was repeated, using the same cleaning solution and equipment, but proceeding through the corridor in the opposite direction.

When dry, the 24 marked tiles of each polish were removed from the floor and transferred to the constant temperature/humidity room for μ_s determinations. The empty tile spaces were filled with (marked) miscellaneous tiles, and the floor was left for two weeks so that the usual traffic density could be re-established.

After traffic returned to normal, the 24 tiles removed for μ_s determinations were pressed back in the floor and the above process, beginning with the application of cleaner solution, was repeated with "Commercial F-1/40" being substituted for NIO-0.5, and then repeated again with "Commercial F-1/20" cleaning solution.

Results and Conclusions

In typical practice, machine scrubbing operations are faster than our test protocol. It is often a two-man operation, with one person applying the cleaner solution immediately around the machine head. This person also takes up the spent cleaning solution after the machine has passed over a section of floor, and he begins the rinse operations. This allows the machine operator to stand on dry floor. We chose to flood the entire floor with cleaning solution, and then allow the solution to remain on the floor until the entire corridor had been scrubbed so that the solution remained on the polish film for a much longer time than would commonly occur. (The machine operator standing on a wet floor is a potentially dangerous situation because of the low slip resistance of the wet floor and the torque exerted by the floor machine.) We intended the longer solution dwell time to exaggerate any effect that the cleaning solution would cause on the μ_s of the films. The use of three machine passes over each tile is also slower and more vigorous than observed with professional custodians. They usually only do two passes, except in those areas that are visibly heavily soiled.

The effect of machine scrubbing on μ_s of the polish films is shown in Table 5. Since the data for the later scrubbings (they are presented in the order of operation) is done on tiles that had already been scrubbed in the earlier operations, the effect shown is cumulative.

There is no change in the slip resistance of the three representative polishes caused by machine scrubbing.

Table 5—μ_s *of polishes after machine scrubbing*[7].

Polish	N-32-25	DP-2-4	Commercial C
Baseline μ_s			
Archive	0.55 +/- 0.01	0.59 +/- 0.03	0.56 +/- 0.04
Determined[a]	0.55 +/- 0.02	0.59 +/- 0.04	0.57 +/- 0.03
After Machine Scrubbing[b]			
NIO-0.5	0.56 +/- 0.02	0.59 +/- 0.02	0.56 +/- 0.03
Commercial F-1/40	0.55 +/- 0.03	0.60 +/- 0.03	0.57 +/- 0.02

Commercial F-1/20	0.56 +/- 0.03	0.59 +/- 0.03	0.58 +/- 0.02

[a] Data is the average of μ_S determination on 4 tiles.

[b] Data is the average of μ_S determinations from 24 tiles.

The Effect of Auto-Scrubbing on μ_s

Auto-scrubbing is a newer process based on a modified floor scrubbing machine that runs in a straight line (much like a lawn mower). The machine applies cleaning solution to the floor, agitates the solution with a brush or scrubbing pad, and then vacuums the spent solution and soils up from the floor. The cleaning solution is pumped onto the floor, usually through the center of the scrubbing pad, and then is immediately collected by a squeegee at the training edge of the pad to be vacuumed into a recovery tank. The typical residence time of cleaning solution on the floor is 1 to 6 seconds.

The majority of autoscrubbers are powered by an on-board bank of batteries that drive the scrubbing head, solution pumps, and wet vacuum (and propulsion motors on large machines). To provide a reasonable battery life and productive work period between battery charges, the downward pressure placed on the scrubbing pad is less than that in a swing scrubbing machine. (This reduced pad pressure also lowers the torque that would divert the machine from a straight line path.)

Relative to swing machine floor scrubbing, auto-scrubbing is much faster, but reduced pad pressure provides less mechanical agitation. Auto-scrubbing is not as efficient at cleaning heavily soiled floors. The speed of auto-scrubbing operations comes from the straight line motion of the machine and automation of the pickup of the cleaning solution.

To compensate for the reduced pad pressure and the reduced residence time of the cleaning solution on the floor, special cleaning formulations are made for auto-scrubbing. These are based on a slightly different class of surfactants that are low foaming. (Foam is a problem in the recovery tank on an auto-scrubber, since it fills with bubbles rather than spent cleaner solution, but still must be emptied when full. Frequent recovery tank discharging operations are expensive because they take time away from productive floor cleaning.) These low foam surfactants are generally not as efficient in attacking heavy oils and greases. Auto-scrub cleaning formulations may also contain solvents to soften oil and grease, making them more easily broken up to be dispersed. These solvents cannot be used in floor cleaner formulations intended for mop cleaning or machine scrubbing because they can also attack the polish film, dulling and damaging it if allowed long residence time.

Unlike the floor cleaning procedures that have been studied thus far, auto-scrubbing does not involve a rinse step. The liquid chemicals applied to the floor are worked, immediately vacuumed up, and the floor is then opened to traffic. The operator of the machine is the first pedestrian on the cleaned floor. A new or reasonably well-maintained auto-scrub machine can be expected to recover from 90 to 95 percent of the liquid that the machine applies[9]. This liquid removal efficiency allows the floor to dry rapidly so that it

[9] Unpublished survey of eight autoscrubbers of assorted sizes from various manufacturers, operated by contract cleaners in the greater Philadelphia, PA, area. Recovery efficiency was determined by weighing the contents of the delivery and recovery tanks before and after machine operation over large, previously cleaned floors.

can be returned to pedestrian traffic with a minimum of "down time." The vacuum efficiency also means that a residue of 5 to 10 percent of the applied chemicals are left on a floor that will soon experience traffic. This potential residue (and the foam problem) dictate low concentrations of the cleaner solutions, to a limit imposed by cleaning efficiency.

With the limitations of lighter abrasion, less efficient dilute chemicals, and shorter residence time, auto-scrubbing is usually considered to be about equal to mop scrubbing in soil-removal efficiency. Its ease of use and low labor cost means that facilities that use auto-scrubbing for routine floor cleaning will perform the operation much more frequently than they would mop cleaning or machine scrubbing. This greater frequency of cleaning results in overall cleaner floors than with these other more effective, but more expensive and less frequent floor cleaning procedure alternatives. Mop cleaning and machine scrubbing have been relegated to spot cleaning or special problems, or facilities that are not amenable to autoscrubbing, such as small floor areas or convoluted walkways.

Polish and Cleaner Selection

The same polish formulations that were used in the mop scrubbing and machine scrubbing evaluations remain representative of the floor polish technologies broadly used in the industry and they are appropriate for this study.

Because our test area is relatively small, we did not have to be overly concerned about the recovery tank foam problem, so we could use the cleaner formulations used in the previous tests, even though we are aware that they are not typically used with this type of equipment and procedure. Since there was no effect on μ_s when these formulations were used with the long residence times and more aggressive scrubbing of the machine scrubbing procedure, our expectation was that any effect on μ_s from autoscrubbing would be due to the residues left by the imperfect recovery vacuum.

Rather than try to learn the art of formulating cleaners for use with an autoscrub, we used an array of commercially available proprietary formulations. These formulations, designated "Commercial B," "Commercial H," and "Commercial M" are reported to be typical of these special cleaning materials. All three proprietary formulations are used at very high dilution (perhaps as a means to control foam in the recovery tank). "Commercial B" and "Commercial M" are designed for use with automatic proportioning/dilution equipment. The manufacturers divulged to us the proprietary dilution ratios engineered into the custom proportioning valves, and we prepared them at this same dilution (by volume) before use. Label directions for "Commercial H" specified a recommended dilution of 1 : 256 (0.5 ounces of formulation per gallon of cleaning solution) for general floor cleaning, and 1 : 128 for heavy duty cleaning. We opted for the higher concentration in order to accentuate any effect the residue would have on μ_s.

Autoscrub Cleaning: Procedure

Though our autoscrubbing operation denied the test floor to pedestrian traffic for only about 15 minutes (scrub time and time to allow any residual solution to dry), we still followed the procedure established for machine scrubbing. This decision was to ensure that the tiles that were selected for μ_s determinations would show the effects of the cleaning operation without being confounded by other effects, such as traffic damage or pedestrian footwear wiping off any residue.

The test area was prepared as described in the procedure for the machine scrubbing tests. In each area, 28 tiles were randomly selected and marked on a corner with a pencil.

Four hours after polish application, four of the marked tiles of each polish were removed for μ_s determinations to provide baseline data. The empty spaces were filled with (marked) miscellaneous tiles.

Cleaner solution NIO-0.5 was charged to the delivery tank of an Advance "Mobilmatic 17" autoscrub machine fitted with a 17 inch (43 cm) diameter 3M red scrubbing pad[8]. The cleaning solution was gravity fed to the floor through the center of the pad at a fixed (valve fully "on") but undetermined rate.

The machine was run in the conventional manner, at a slow walking speed.

When the entire corridor had been scrubbed, it was allowed one hour to dry and the above procedure was repeated, using the same cleaning solution and the same equipment, but passing over the floor in the opposite direction.

After the second scrubbing, the remaining 24 marked tiles of each polish were removed from the floor and set aside. The tile spaces were filled with other miscellaneous tiles. The delivery tank and recovery tank on the machine were drained and flushed.

The above was repeated each day, after the 24 marked tiles of each polish had been returned to the floor. After 20 autoscrubbing operations had been completed, the marked tiles were removed and transferred to the constant temperature/humidity room for μ_s determinations. The empty tile spaces were filled with (marked) miscellaneous tiles.

The above procedure was repeated, beginning with stripping the floor, but the five test cleaning solutions were sequentially substituted for NIO-0.5.

Results and Conclusions

The test with "Commercial F-1/20" as the cleaner solution was aborted because the recovery tank filled with foam before the first auto-scrubbing was completed. The machine shuts off the wet vacuum when the recovery tank is sensed to be full.

The μ_s data for the other cleaner solutions is shown in Table 6. Because each test of a cleaning formulation began with a freshly stripped and polished floor, the μ_s data after autoscrubbing is *not* cumulative. That is, any change in μ_s is a consequence of auto-scrubbing with only that cleaner formulation.

Within the statistical limits of the data, auto-scrubbing with any of the selected cleaner formulations, even those designed for use in a cleaning procedure that includes a rinse step, had no effect on the slip resistance of the tested polishes.

Table 6–μ_s *of polishes after autoscrubbing*[7].

Polish	N-32-25	DP-2-4	Commercial C
Baseline μ_s			
Archive	0.55 +/- 0.01	0.59 +/- 0.03	0.56 +/- 0.04
Autoscrubbing Formulation			

NIO-0.5			
Baseline[a]	0.55 +/- 0.02	0.59 +/- 0.04	0.57 +/- 0.03
After 20 scrubs[b]	0.53 +/- 0.03	0.56 +/- 0.03	0.56 +/- 0.03
Commercial F-1/40			
Baseline[a]	0.53 +/- 0.02	0.57 +/- 0.04	0.55 +/- 0.03
After 20 scrubs[b]	0.54 +/- 0.03	0.55 +/- 0.03	0.55 +/- 0.02
Commercial F-1/20			
Baseline[a]	0.55 +/- 0.02	0.59 +/- 0.04	0.57 +/- 0.03
After 20 scrubs[b]	Aborted for excessive foam in recovery tank		
Commercial B			
Baseline*	0.56 +/- 0.02	0.60 +/- 0.04	0.56 +/- 0.04
After 20 scrubs**	0.54 +/- 0.03	0.59 +/- 0.03	0.56 +/- 0.03
Commercial H			
Baseline[a]	0.55 +/- 0.02	0.59 +/- 0.04	0.57 +/- 0.03
After 20 scrubs[b]	0.52 +/- 0.02	0.59 +/- 0.02	0.53 +/- 0.03
Commercial M			
Baseline[a]	0.55 +/- 0.02	0.59 +/- 0.04	0.56 +/- 0.03
After 20 scrubs[b]	0.57 +/- 0.02	0.61 +/- 0.02	0.58 +/- 0.03

[a] Data is the average of μ_S determination on four tiles.

[b] Data is the average of μ_S determinations from 24 tiles (96 drops).

The Effect of Traffic on Polish μ_s

The common procedures in routine floor polish maintenance, damp mopping, mop scrubbing, machine scrubbing, and auto-scrubbing, have been shown above to have no effect on the slip resistance of polishes. This circumstance allows us to test the effect of pedestrian traffic and wear on the slip resistance of polishes.

We delayed testing the effect of traffic on μ_s until we had established whether the cleaning operations had an effect because appropriate reflection, and discussions with numerous custodians and contract cleaners, convinced us that traffic is always accompanied by some form of cleaning maintenance.

However, immediately after a polish is applied to a floor it gets trafficked with no wet maintenance until the cleaning schedule is put into effect. Thus, for a (usually) short time the polish is trafficked without cleaning. (No one bothers to clean a newly applied polish that is not yet soiled.) This period of trafficking without wet cleaning maintenance is a small fraction of the working life of a finish.

Recognizing that polish films subjected to moderate-to-heavy pedestrian traffic without routine cleaning maintenance is not a typical occurrence, we undertook the study anyway. Our only concession to maintenance was cleaning loose soil from the tiles before μ_s determinations. This was done so that the variable in the frictional interface being tested would be the polish film surface, rather than movable loose soil.

Polish Selection

The polish formulations that we have been working with in the previous sections of this study are designed for high traffic with minimal maintenance, and they are representative of the most widely used technologies in the industrial and institutional polish industry. We again used them in this phase of the study.

No other chemicals were used in this part of the study.

Procedure

The entire corridor was stripped of polish as has been described earlier[4]. The wear test area was divided into three equal 12 foot (3.66 m) lengths.

Three coats of the designated polishes were mop-applied to each area, with 45 minutes drying time allowed between coats. After the last coat was applied, the floor was left to dry overnight. The polishes applied to the ends of the wear test area (DP-2-4 near the fire doors, N-32-25 near the cross-corridor) were also applied into the areas with different traffic patterns (see Fig. 3). The application of each polish was also extended to the rows of tiles abutting the corridor walls so that the entire corridor was polished.

After drying overnight (approximately 5 hours), the corridor was opened to pedestrian traffic.

No maintenance was done to the corridor during the test period other than picking up the odd piece of rubbish that could be considered a slip hazard. (The corridor was checked for loose litter a couple of times each day.)

When the gloss of the N-32-25 area had dropped to about half its initial gloss (visual evaluation only), 24 tiles were removed from each polish area and marked as to their location in the test. Tiles selected for each polish were the 12 tiles of the middle row and 6 tiles from each of the two rows adjacent to the central row since these three rows had been established as receiving the most traffic.

The marked tiles were lightly wiped with dry, clean, unsized cheesecloth to remove any loose dust and soil. They were then transferred to the constant temperature/humidity room for μ_s determinations.

Results and Conclusions

The data from this wear test are presented in Table 7.

There is a possibility that the floor was swept (untreated cotton dust mop) at least once during the first week of the test and again once early in the second week. This was denied by the crew leader of the custodial contract cleaning service responsible for the building. In the middle of the second week of this test, a very small pile of dirt was placed on the tiles adjacent to the wall and next to the end of the swing of the fire doors. This was outside the traffic lanes and it remained undisturbed throughout the remainder of the test, indicating that the custodial staff was not surreptitiously cleaning this "no maintenance" test floor (or that they were very inefficient).

Table 7–μ_s *of polishes after traffic (no maintenance)*[7].

Polish	N-32-25	Commercial C	DP-2-4
Baseline μ_s			
Archive	0.55 +/- 0.01	0.56 +/- 0.04	0.59 +/- 0.03
After 6 weeks of traffic			
Central Row[a]	0.48 +/- 0.05	0.45 +/- 0.07	0.54 +/- 0.05
Adjacent Rows[a]	0.51 +/- 0.04	0.49 +/- 0.06	0.56 +/- 0.05
Overall[b]	0.50 +/- 0.05	0.46 +/- 0.07	0.55 +/- 0.05

[a] Data is the average of μ_S determinations from 12 tiles and is included in the overall data.
[b] Data is the average of μ_S determinations from the above 24 tiles considered as one sampling.

Because it bears on the following discussion, the subjective polish performance ratings at the end of the wear test are shown in Table 8.

A comparison of the tiles removed for μ_s determinations with those of the same polish along the corridor wall (which receive essentially no traffic) showed the selected tiles to be heavily scuffed and soiled. The tiles from the "Commercial C" and N-32-25 areas were also black marked. All of the test area tiles were lower in gloss than the tiles along the walls, though those from DP-2-4 had retained more of their initial gloss and scuffed less that the other two polishes.

Soiling in all of the polishes was seen as a general dingy gray or yellowish-gray coloration. Soiling was more evident in the tiles from the central row of each polish for both "Commercial C" and N-32-25. The DP-2-4 coated tiles were less soiled than the other two polishes, and the central row tiles were less noticeably soiled than the adjacent rows, though the central row tiles were more heavily scuffed and had lower gloss than the adjacent rows of DP-2-4.

Table 8–*Relative performance of polishes after traffic (no maintenance).*

Polish	N-32-25	DP-2-4	Commercial C
Gloss Retention[a]			
Central Row	4.0	5.0	4.0
Adjacent Rows	4.5	6.0	5.0
Overall[b]	5.5	6.5	5.5
Soil Resistance[a]			

Central Row	4.0	7.0	4.5
Adjacent Rows	4.5	7.0	5.0
Overall[b]	5.5	7.5	5.5
Scuff Resistance[a]			
Central Row	5.0	6.5	5.5
Adjacent Rows	5.5	7.0	6.0
Overall[b]	6.5	8.0	6.5
Black Heel Mark Resistance[a]			
Central Row	6.0	10	7.0
Adjacent Rows	7.0	10	7.5
Overall[b]	7.5	10	8.0

[a] Subjective evaluations, on a scale of 1 to 10 where high values are preferred performance. These numbers may be translated to subjective ratings as follows: 10 = Excellent; 8 = Very Good; 6 = Good; 4 = Fair; 2 = Poor.
[b] Overall evaluations are of the entire 5' x 12' (1.52 m x 3.66 m) segment of the wear test area and are not limited to the tiles in the central and adjacent rows.

No effort was made to avoid areas of the tiles that had been black marked or heavily scuffed in the μ_s testing. To do so would have not only slowed the tedious μ_s determinations, but could also skew the data, because pedestrians usually do not avoid walking on scuff or black heel marks.

The data indicates that pedestrian traffic with no cleaning maintenance reduces the slip resistance of "Commercial C" and N-32-25, though the slip resistance of the films did not drop below the $0.5\mu_s$ conformance criteria. (ASTM D2047 specifies that μ_s is to be rounded to one decimal place for comparison with the conformance criteria.) The reduced μ_s conclusion is only weakly justified by the statistics for the μ_s determinations on "Commercial C", and an even weaker justification is present for N-32-25 because of the wide scatter (large standard deviation) in the data. DP-2-4 did not show a drop in slip resistance in this test.

The drop in slip resistance is thought to be due to ingrained soil. Ingrained soil would change the variable frictional surface from the polish film to some hybrid film of dirt/polish that may have a lower μ_s. It is known that loose soil on a polish film will lower slip resistance quite dramatically, but this has always been presumed to be due to the mobile nature of the loose dirt, coupled with the hardness of the polish film. Loose soil acts as a lubricant on the polish surface. "Commercial C" and N-32-25 both showed heavy soil impregnation and also a drop in μ_s.

The reduction in polish slip resistance could also be a function of black marking, since "Commercial C" and N-32-25 both showed heavy black marking and lower μ_s while DP-2-4 shows neither. Heel marks are often thought of as a special form of ingrained soil. Like ingrained soil, black heel marks are an emulsion of a solid (pigmented rubber or soil particles) in the solid floor polish film that serves to stiffen the polish film.

The reduction in polish slip resistance could also be a function of scuffing. All of the polishes showed scuffing, and scuffing was more severe in the central row than in the adjacent rows. The central row tiles showed the greater drop in μ_s. A scuff mark is known to be a disruption of the polish film surface, and it is possible that this disruption exposes internal film chemical constituents that have a different (lower) μ_s.

The choice between these possible causes of the drop in μ_s can be resolved by introducing maintenance procedures which reduce or eliminate the suspects. The easiest one to eliminate is ingrained soil, because that requires only that we repeat the wear test but with routine cleaning maintenance.

The Effect of Traffic and Routine Cleaning Maintenance on Polish μ_s

Thus far, we have tested the individual components of the environment of a polish in traffic. Though the testing has been done as though each operation were independent of the others, in the real world of polish films, they are combined to make up the maintenance program. Janitors do not apply polish, damp mop, mop scrub, machine scrub, or autoscrub floors that do not experience traffic. It is also rare (but not impossible) for polish films to experience traffic without any maintenance. Cleaning maintenance begins soon after the polished floor is opened to traffic. This routine maintenance prevents the polish film from getting into a worn condition such that drastic measures must be taken to restore the appearance of the floor. Because of the time and labor costs involved in these drastic restorative processes (which include deep scrubbing and top coating, or complete stripping and replacement of the finish) it is much less expensive to do routine maintenance. The frequency of routine cleaning maintenance is ideally determined by the traffic density and how much soil is tracked onto the floor by the pedestrians. In practice, the frequency of maintenance is set by the budget available to the facilities manager and a schedule that provides maintenance to an area as a function of the peak traffic density and soiling.

The common experience is that cleaning maintenance and pedestrian traffic are coupled, and so we combined them in this stage of the testing.

Polish and Cleaner Selection

The three polishes used in the previous parts of this testing are appropriate for this one as well.

Cleaner formulation "Commercial H," used in the study to determine the effect of auto-scrub cleaning on μ_s, is used by the contract cleaner to maintain the remainder of the facility. It was selected for use in their auto-scrub equipment because it demonstrated efficient cleaning ability against the types and amount of soils normally encountered on the resilient floors on site. With this recommendation (and the donation of material to use in our testing), we selected "Commercial H" for routine auto-scrub maintenance of the traffic wear test. The formulation was used at a 1: 256 dilution in accord with the label recommendation for general floor cleaning. The maintenance schedule for the wear test was the same as that used by the contract cleaner on the cross-corridor intersected by the test area.

Procedure

The entire corridor was stripped out[4], and the wear test area was prepared in a manner identical to the previous test of the effect of traffic without cleaning maintenance. Three coats of each polish were applied with string mops.

After allowing the polishes to dry overnight (approximately 6 hours), the corridor was opened to pedestrian traffic.

The corridor was swept daily with a clean, untreated, cotton dust mop.

Twice weekly the corridor was autoscrubbed with an Advance "Mobilmatic 17" autoscrub machine fitted with a 17 inch (43 cm) diameter 3M red scrubbing pad and "Commercial H" cleaning solution prepared at a 1:256 dilution (by volume).

The gloss of the central row of tiles for each polish was measured with a 60° gloss meter once each week, immediately after the floor had been auto-scrubbed. The test was run until the average gloss of the central tile row of any polish dropped to half of the average gloss reading taken after the first day of the test. The polishes were also subjectively evaluated for visual gloss retention (by comparing with the tiles along the wall), scuff resistance, mark resistance, and soil resistance.

At the end of the test, the corridor was again autoscrubbed and, after allowing one hour to dry, in each polish area the 12 tiles of the central row and 6 tiles from each of the two rows adjacent to the central row were taken up and marked. The 24 tiles of each polish were transferred to the constant temperature/humidity room for μ_s determinations.

Results and Conclusions

The μ_s data from this wear test are presented in Table 9 and the polish performance data is in Table 10.

Table 9–μ_s *of auto-scrubbed polishes in traffic*[7].

Polish	N-32-25	DP-2-4	Commercial C
Baseline μ_s			
Archive	0.55 +/- 0.01	0.59 +/- 0.03	0.56 +/- 0.04
After 12 weeks of traffic			
Central Row[a]	0.53 +/- 0.04	0.61 +/- 0.04	0.54 +/- 0.04
Adjacent Rows[a]	0.53 +/- 0.03	0.60 +/- 0.04	0.55 +/- 0.05
Overall[b]	0.53 +/- 0.03	0.61 +/- 0.04	0.54 +/- 0.04

[a] Data is the average of μ_s determinations from 12 tiles and is included in the overall data.

[b] Data is the average of μ_s determinations from the above 24 tiles considered as one sampling.

Table 10–*Relative performance of auto-scrubbed polishes in traffic.*

Polish	N-32-25	DP-2-4	Commercial C
60° Gloss			
Initial	76	82	78
Final	35	48	41
Gloss Retention[a]			
Final	5.5	6.5	5.5
Soil Resistance[a]			
Final	8.0	8.5	8.0
Scuff Resistance[a]			
Final	5.0	6.5	5.0
Black Heel Mark Resistance[a]			
Final	6.0	9.5	6.0

[a] Subjective evaluations, on a scale of 1 to 10 where high values are preferred performance. These numbers may be translated to subjective ratings as follows: 10 = Excellent; 8 = Very Good; 6 = Good; 4 = Fair; 2 = Poor.

Formulation N-32-25 was the first polish to drop to half of its initial gloss, triggering the termination of this test. The duration of the test was slightly more than twice that of the previous test with no cleaning maintenance, confirming the well-accepted industry lore that clean floors maintain their gloss better than soiled floors.

The slip resistance of all of the tested polishes is not changed by the combination of pedestrian traffic and cleaning maintenance. Though the test was run twice as long as the traffic test with no maintenance, the amount of imbedded soil in the test polishes was considerably less. Similarly, the values of μ_s remained unchanged. This result implicates the soil impregnation in the films as the cause for the reduction in slip resistance noted earlier.

The amount of scuffing and black heel marking in this test was greater than observed in the previous test, consistent with the longer time in traffic, but the μ_s was not decreased. Traffic damage in the form of black marking and scuffing are not responsible for the decrease in slip resistance observed earlier.

Since the changes in polish films that result from traffic damage, including scuffing, black heel marking, and the accumulation of micro-scratches that result in the gradual loss of gloss, do not reduce the slip resistance of polishes in traffic, we can now test the effect of other maintenance procedures on μ_s. The common maintenance procedures, known to modify the polish surface, are spray buffing and high speed burnishing.

Effect of Abrasive Machine Maintenance on Polish μ_s

Spray buffing (SB) maintenance and high speed burnishing (HSB) maintenance are both known [4,5] to be abrasive processes that remove the surface of a polish film. They are highly effecting in repairing traffic damage, which is largely limited to the surface of the film, by abrasively removing the very top of the surface and abrasively smoothing the surface. The result of these maintenance procedures is repair of traffic damage in the form of micro-scratches, cuts and scuffs, and restored or improved gloss.

Spray buffing and high speed burnishing differ in the equipment used, the techniques employed, the auxiliary chemicals, and the demands they place on the polish film. We first consider spray buffing, the oldest of these new maintenance procedures.

Spray Buffing Maintenance

Spray buffing was invented by custodians who found that when a wet polish formulation was sprayed or misted on a traffic-damaged floor and then immediately buffed with a conventional 175 rpm swing floor machine, the gloss of the worn polish film was improved. Both black marks and scuffs were removed. It was also noted that the spray buffing process cleaned the floor, with the dirt becoming imbedded into the scrubbing pad. The procedure improved the overall appearance of the floor polish and lengthened the time between expensive stripping operations. These were sufficient reasons for the procedure to find moderately broad acceptance, but SB maintenance required the development of new skills in the operation of the machine, and until these were mastered by the custodian there was a danger that the polish film would be wrecked.

The problem of the learning curve for the SB technique was addressed by machine manufacturers who first developed higher speed (300 and 400 rpm) swing machines. These machines were only marginally successful because every facility already had a (paid for) 175rpm swing machine that worked just about as well.

Eventually higher speed machines were built (1 000 to 1 500 rpm) that got around the problems of spray buffing technique by operating in a straight line, similar to an auto-scrubber or a lawn mower. The higher rotational speeds were attained by having the machine do less abrasive work (less pressure on the pad or only a small part of the pad in contact with the floor). These machines were able to spray buff as much as 10 000 square feet (929.0 m^2) per hour, compared to 1 500 square feet (139.36 m^2) per hour for swing machines.

The polishes used in SB maintenance are those that were available at the time the process was being developed. That is, they are the conventional minimal maintenance polishes similar to the polish formulations we have been using in these studies. The subsequent development of polish formulations specifically intended for SB maintenance produced marketing failures. However, alternative formulations, called "Spray Buffing Media (SBM's)" to replace and improve on the spray-applied polish, were quite successful. They were easier and cheaper to use, did a better job in cleaning the floor of soils, did not cause as much (non-dirt) solids to pack into the scrubbing pad, and made the pad easier to clean. The shorter dwell time of the higher speed machines over any given tile meant that special spray buff media had to be formulated to work with these faster machines.

The productivity advantages of the higher speed, straight-line machines means that the swing machine spray buffing technique is currently a spot-cleaning and repair technique or relegated to those facilities that are relatively small or have convoluted aisle ways. That is, it is now largely relegated to those types of facilities that still do swing machine scrubbing. Though not as extensively used as it once was, SB maintenance is still the maintenance procedure of choice in some facilities, so we now consider its effect on polish slip resistance.

Chemicals Selection

Spray buffing maintenance utilizes conventional, minimal maintenance floor polishes similar to the three polish formulations that have been used in the previous phases of this work.

Spray buff medium formulation SBM-44-1 is representative of the relatively simple formulations used in the industry for swing machine spray buffing maintenance. We purchased two commercial formulations to use in this phase of the study, but analysis showed them to be identical to SBM-44-1 (except for the addition of a very small amount water soluble dye in one case, and an opacifier in the other case). The formulation can be made as a concentrate (for dilution by the custodian), but the formulation details given in the Appendix are for the material as it is to be used.

For high speed (straight line) spray buffing, spray buff medium formulation SBM-44-2 was found to be typical of two of the three commercial formulations that we purchased. The third formulation, "Commercial T", differed in having higher solids, a different alkali soluble resin, solvents, and waxes and it included a dye and opacifier. This proprietary, commercial formulation was used in this testing after we were told by the manufacturer that it was the largest selling high speed spray buffing medium in the New England and Mid-Atlantic states. It was used at the recommended dilution ratio of 1:3.

It had previously been established[10] by ESCA and reflectance infrared spectroscopy that there is no SBM residue on the polish surface after SB maintenance. Any change in slip resistance observed as a result of spray buffing maintenance will be due to mechanical abrasion of the film surface exposing underlying chemical structure.

Procedure

The procedure used is very similar to that of ASTM D3758.

The entire corridor was stripped out[4] and the wear test area was prepared in a manner identical to the previous test of the effect of traffic without cleaning maintenance. Three coats of each polish were applied with string mops, with DP-2-4 applied near fire doors the and "Commercial C" applied near the cross-corridor.

After allowing the polishes to dry overnight (approximately 6 hours), the corridor was opened to pedestrian traffic.

The corridor was swept daily with a clean, untreated, cotton dust mop.

---

[10] Unpublished report by Dr. Walter Platek, Rohm and Haas Company, Spring House, PA.

Every other workday the entire corridor was spray buffed with a 19 inch (48 cm) diameter Hako 330 rpm swing floor machine fitted with a red scrubbing pad[8]. SBM-44-1 was used as the spray buffing medium. The spray buffing medium was squirted on the floor at such a rate that the trailing edge of the swing showed a damp streak no more than one inch (25 mm) wide. A different, clean scrubbing pad was used for each test polish (the same pad was kept for use on each polish throughout the duration of this test). Since spray buffing maintenance is known to be a cleaning procedure, as well as repair, no other cleaning was done on the test floor.

After eight weeks (20 spray buffing operations) the test was terminated and the 12 central row of tiles from each polish were removed, as were 6 additional tiles from each of the two rows adjacent to the central row. These 24 tiles from each polish area were transferred to the constant temperature/humidity room, and the μ_s of the polishes was measured.

The tiles were returned to the floor, and the test begun again, with straight line, high speed spray buffing maintenance.

The test area was visually divided in half along its length (middle of the central row of tiles). The south side was maintained with spray buffing medium formulation SBM-44-2, and the north side with spray buffing medium "Commercial T."

High speed spray buffing was done with an NSS 21 inch (53 cm) diameter 1070rpm straight line high speed buffing machine. A cream-colored high speed buffing pad[11] (slightly less aggressive than a red scrubbing pad) was used for each polish and spray buffing medium combination.

Our high speed spray buffing machine does not have an automatic sprayer on the leading edge of the pad housing, so the procedure became a two-person operation. The extra person sprayed a periodic stream of the spray buffing medium in front of the rapidly moving machine. Each polish was spray buffed twice in each operation.

After eight weeks (40 high speed spray buffing operations) the 24 tiles in the first and second rows adjacent to the central row for each polish were removed for μ_s determinations. The central row was not removed because it had been spray buffed with both of the test spray buffing media and the overlap was uncertain.

Results and Conclusions

The μ_s data from this test are presented in Table 11 and the polish performance data is in Table 12.

Table 11–μ_s *of polishes after spray buffing*[7].

Polish	N-32-25	DP-2-4	Commercial C
Baseline μ_s			
Archive	0.55 +/- 0.01	0.59 +/- 0.03	0.56 +/- 0.04

[11] Available from Etc., Inc., Hendersonville, SC; sold only through janitorial supply distributors.

Low Speed (swing) spray buffing maintenance with SBM-44-1			
20 operations[a]	0.54 +/- 0.02	0.58 +/- 0.03	0.56 +/- 0.03
High Speed (straight line) spray buffing maintenance with SBM-44-2			
40 operations[a]	0.55 +/- 0.03	0.59 +/- 0.03	0.55 +/- 0.02
High Speed (straight line) spray buffing maintenance with "Commercial T"			
40 operations[a]	0.57 +/- 0.03	0.59 +/- 0.03	0.57 +/- 0.03

[a] Data is the average of μ_S determinations from 24 tiles.

The frequency of spray buffing operations was 2.5 to 5 times that which would be typically used for a passage with the amount of traffic and soil as our wear test area. We were not concerned with the cost of maintenance, and wished to exaggerate the effects, if any, of the procedure.

As shown in the data of Table 11, neither low speed nor high speed spray buffing has any effect on the slip resistance of the polishes tested.

A comparison of the chemically different high speed spray buffing media shows that the composition of the media, at least within the range of compositions represented by these materials, has no effect on the slip resistance of the polishes.

As a routine exercise in conducting wear tests, the polishes were evaluated for various performance factors before and after each maintenance operation. (It costs very little extra to get every bit of possible information, including data that is not pertinent to the objective of the study.) It was noted immediately after low speed spray buffing that the subjective "Paper-Under-Foot" test for slip resistance [6] indicated that the test polishes would show reduced μ_s. This was not true for the polish films after high speed spray buffing. We were thus surprised when the μ_s data showed that there was no change in slip resistance after low speed spray buffing maintenance. Though it is a subjective test, we have found the "Paper-Under-Foot" test to be both reliable and highly sensitive to changes in slip resistance.

Table 12–*Relative performance of spray buffed polishes.*

Polish	N-32-25	DP-2-4	Commercial C
Gloss Retention[a]			
Final SBM-44-1	8.0	8.5	8.0
Final SBM-44-2	8.5	8.5	9.0
Final "Commercial T"	8.5	8.5	8.5
Soil Resistance[a]			
Final SBM-44-1	9.0	9.5	9.0

Final SBM-44-2	7.5	8.5	8.5
Final "Commercial T"	7.5	8.8	8.5
Scuff Resistance[a]			
Final SBM-44-1	8.0	9.5	7.5
Final SBM-44-2	7.5	9.5	8.0
Final "Commercial T"	7.0	9.5	8.0
Black Heel Mark Resistance[a]			
Final SBM-44-1	8.0	10.0	8.0
Final SBM-44-2	7.0	10.0	8.0
Final "Commercial T"	7.5	10.0	8.5

[a] Subjective evaluations, on a scale of 1 to 10 where high values are preferred performance. These numbers may be translated to subjective ratings as follows: 10 = Excellent; 8 = Very Good; 6 = Good; 4 = Fair; 2 = Poor.

Preliminary low and high speed spray buffing experiments showed that low speed spray buffing lowers the μ_s of a polish film, but the coefficient rapidly and spontaneously reverts back to the original μ_s value.

Equilibration of μ_s after Low Speed Spray Buffing Maintenance

To test for changes in μ_s very soon after spray buffing we modified the protocol so that μ_s determinations could be made almost immediately after the maintenance procedure was completed.

As mentioned previously, there are no known laboratory procedures that mimic the effects of large scale floor maintenance procedures, so spray buffing still had to be done on the test floor. Determinations of μ_s by the procedure of ASTM D2047 require 3 to 5 minutes per tile, even for a very adept operator. When the time between spray buffing and determination of μ_s is important there is little point in spray buffing a large floor. Instead, a set consisting of only four tiles was used, and the test replicated a number of times to obtain a statistically meaningful sample of data on μ_s as a function of time after spray buffing.

Chemical Selection

Because the phenomenon of time-dependent μ_s change was first observed with polishes N-32-25 and DP-2-4, these formulations were used in this study. Subsequent, independent testing showed that the magnitude of the change in μ_s is greatest for polishes based on the same polymer and wax blend as DP-2-4, but the phenomenon appears to be common to some extent in all of the polishes that were screened. The absolute amount of drop in μ_s, or the drop in μ_s as a proportion of the fully equilibrated μ_s, is highly variable.

High speed spray buffing did not show the time-dependent change in μ_s, but the high speed experiment also used a different spray buffing medium, SBM-44-2. Though formulation SBM-44-2 is not suited for use in a low speed spray buffing operation, it was included in this testing to determine if the lack of μ_s change is due to the auxillary chemicals employed in the high speed spray buffing process.

Water also works as a spray buffing medium, though not as well as formulations that also contain other lubricating materials [4]. Water was included in this study because the simplicity of the formulation (only deionized water) could clarify possible chemical interactions as the cause of time-dependent μ_s changes.

Procedure

Four tiles were dip coated per ASTM D4103 to apply four coats of DP-2-4. After drying overnight, these tiles were inserted into the center of the wear test area. They were spray buffed as described earlier (330 rpm swing machine, red scrubbing pad[8]) with SBM-44-1 as the spray buffing medium.

Immediately after spray buffing a timer was started and the tiles were transferred to the constant temperature/humidity room for μ_s determinations. As each data point was generated, the time lapse was noted.

When all four tiles had been measured for μ_s, they were shuffled and replaced in the floor for a replication of the above testing. Eight replications were run.

This protocol was repeated with deionized water as the spray buffing medium, and then with SBM-44-2 as the spray buffing medium.

The entire process, including the alternative spray buffing media, was then repeated with polish N-32-25.

Results and Conclusions

Rather than a data table, the result are shown more effectively in Figs. 4 and 5, plots of μ_s as a function of the time lapse from the completion of the spray buffing operation for DP-2-4 and N-32-25. The earliest μ_s measurements were made at 1.25 minutes. Time lapses greater than 8 minutes gave μ_s values that were essentially the same as the archive data.

The data and the curves for both polishes and all three spray buffing media describe a zero order reaction or process (determined by least-squares fitting of the raw data to different order rate laws). Extrapolating the rate equation back to time = zero (the instant spray buffing was completed) yields the lowest value of μ_s. In none of the polish/media combinations studied does this go below 0.50.

The drop in μ_s is greatest for SBM-44-1 and least for water as the spray buffing medium, with SBM-44-2 being intermediate to these, for both polishes. It is tempting to assign the drop to either some constituent in the alkali soluble resin in the two spray buffing media, or to an interaction between the polish and this hypothetical constituent, because SBM-44-1 has about twice the amount of alkali soluble resin as does SBM-44-2.

This argument fails when it is noted that water caused a drop in μ_s as well. The alkali soluble resin is not volatile, so if it were the cause of the drop there is no mechanism for restoring μ_s. It is known[10] that solid materials from spray buffing media are not left on the film as a residue.

A more convincing argument can be made that the drop in μ_s is due to water absorbed into the surface of the polish film. Of course, water is present in all three of the spray buffing media, but SBM-44-1 also contains a hydroxyalkyl amine (low volatility and low odor), and SBM-44-2 contains ammonia. The amines are in the formulations so that the alkali soluble resin will remain in solution (where it acts as a lubricant between the pad and film) and to neutralize the acid functionality in the film (amines unlock some of the metal crosslinking, making the film softer, less abrasion resistant, and more hydrophilic). The amines cause the film to absorb additional water from the spray buffing medium. Some water can still be absorbed if no amine is present (depending on the hydrophilicity of the film), but not as much as when some surface acids are uncrosslinked and made available. Ammonia, being more volatile leaves the film quickly, allowing the crosslinking to rapidly recover, and residual water is lost as the film returns to being more hydrophobic. The same process occurs more slowly with the less volatile alkyl amine of SBM-44-1.

This explanation is supported by the observation that DP-2-4 films, which are inherently more acid-rich and hydrophilic, showed a larger drop in μ_s than did N-32-25 films. A greater drop in μ_s for DP-2-4 was evident with water as well as the different spray buffing media formulations and their different amines.

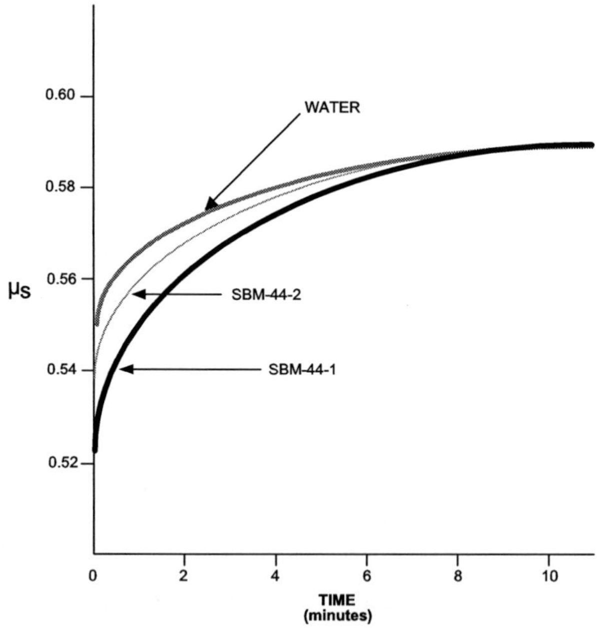

Fig. 4—*Time-dependent change of* μ_s *for DP-2-4.*

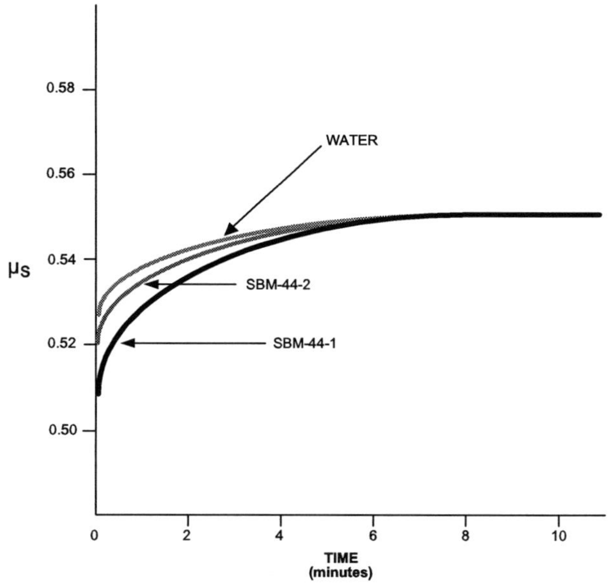

Fig. 5—*Time-dependent change of* μ_s *for N-32-25.*

The evaporation of water from a large surface area is a zero order kinetic process, a function of (fixed) surface area and ambient conditions.

When the μ_s of the films were measured, the films were dry to touch. The drop in μ_s is not due to the water acting as a lubricant or becoming the variable surface in the frictional interface. Rather, the μ_s values measured immediately after spray buffing are the μ_s of a solid, though slightly swollen, hydrated film.

There is no convenient way to measure the hydrophilicity of a film, but it can be inferred from indirect evidence. For instance, more hydrophilic polish films are known to have better black heel mark resistance. When a series of polishes was ranked for black heel mark resistance the ranking was almost identical to a ranking of the magnitude of the drop in μ_s shown in spray buffing. (The discrepancies in the rankings may be due to statistical noise in the μ_s data, since the screening experiment was not highly replicated.)

High Speed Burnish Maintenance

High speed burnishing is the result of machine manufacturers' efforts to speed the spray buffing process and remove the sensitivity of the results to the operator's skills. Manufacturers designed machines like those described above for high speed spray buffing, but with rotary speeds of 1000 to 4000 rpm. It was found [4,5] that at speeds above 1100 to 1500 the presence or absence of a spray buffing medium had no discernable effect on the results produced, so high speed burnishing is a dry process, with no auxilary chemicals involved in the procedure.

High speed burnishing is an abrasive process [5] that removes a very thin layer of the floor finish, and with it the scuffs and small scratches from traffic that have degraded appearance. The process also smooths the polish surface so that gloss is restored and even improved. The gloss from high speed burnishing is a result of abrasive smoothing of the film and is a function of the circumferal speed (diameter x rpm) of the pad. Higher rotational speeds produce higher gloss, reaching a maximum at about 2500 rpm.

The ability to repair traffic damage by high speed burnishing is a function of the abrasive work done by the machine and pad on the floor polish. In general, this is dependent upon pad pressure. For plug-in electric machines the amount of energy available to drive the pad is limited by what can be transmitted through the building circuit breakers and the long cord to the machine. Battery-driven electric machines are limited by the time that a charge will last. Driving the pad at higher speed consumes more energy, so with a restricted power source the amount of energy remaining to do any abrasive work on the polish is reduced.

The power limitation was addressed with internal combustion (usually propane fueled) driven burnishing pads. These machines are complicated, expensive, and have exhaust emission problems in an internal environment.

The speed / abrasive power compromise is also a gloss / repair compromise. It has been resolved to the extent that most high speed burnishers operate at about 2000 rpm.

Relative to spray buffing, high speed burnishing is much more efficient at restoring, and even improving, the initial gloss of a trafficked polish. High speed burnishing is much poorer at repairing the traffic damage, such as scuffs, light cuts and scratches, and (with some work) black marks. High speed burnishing is not considered a cleaning operation, so the floor must be cleaned in a separate operation. The mechanics of high speed burnishing mean that surface soils that are not removed before the procedure may actually be driven into the film, so the floor *must* be cleaned before burnishing.

Formulators responded to the poor repairability of high speed burnishing by developing formulations that had poorer abrasion resistance (they usually made the film softer) so that even the reduced abrasion of the high speed machine would remove the polish surface, and with it the traffic damage. Of course, reduced abrasion resistance means that the film is also less resistant to the abrasion of traffic, so these special formulations deteriorate more rapidly in traffic.

High speed burnishing, which is less efficient at traffic repair than spray buffing, results in improved overall polish appearance because its speed and ease of use allows it to be performed much more frequently. This is analogous to auto-scrubbing, which is less efficient at cleaning than mop scrubbing or machine scrubbing, but results in cleaner floors because its speed and ease of use allow it to be performed more frequently. Daily burnishing of high traffic density floors is not unusual, just as daily auto-scrub cleaning is common and effective in high traffic areas.

Polish Selection

Though minimal maintenance polishes such as we have been using in the previous parts of this research are subjected to high speed burnishing, it is more common that formulations specifically designed for high speed burnishing are used with this procedure.

As representative of minimum maintenance polishes we included formulation DP-2-4 in this work.

As representative of a high speed burnishing polish we used formulation HS-04-9. This is a softer, less abrasion resistant, very hydrophilic, conventionally metal crosslinked, styrenated acrylic polish. Formulation details for HS-04-9 are in the Appendix. At the time this research was initiated, the technology of HS-04-9 was the fourth broadest used floor polish technology in the world. There is also an archive of μ_s measurements for HS-04-9, and the archive average is based on 2448 determinations.

A proprietary commercial formulation, "Commercial E", was also included in this study because it was reported to be the largest selling high speed burnishing polish formulation in North America. It is a much softer, lower abrasion resistant polish than HS-04-9, with lower metal crosslinking and a more hydrophobic styrenated acrylic polymer.

Procedure

The wear test was prepared as described previously for the test of spray buffing maintenance, and five coats of each polish were applied to thirds of the wear test area. After drying overnight (about 4 hours) the wear test was opened to pedestrian traffic.

The wear test was dust mopped and then auto-scrubbed each day with "Commercial H," as described previously.

Immediately after auto-scrubbing, the entire wear test area was high speed burnished daily with an NSS 21 inch (53 cm) diameter 1 070 rpm straight line electric high speed buffing machine fitted with a synthetic cream-colored high speed burnishing pad[11].

After four weeks of traffic and daily high speed burnishing, the 12 tiles of the central row for each polish and 12 tiles from the two rows adjacent to the central row were removed and transferred to the constant temperature/humidity room for μ_s determinations per ASTM D 2047.

After μ_s determinations were made, the tiles were returned to the wear test and the auto-scrubbing and high speed burnishing was continued. The machine used for this stage of testing was a Hako 21 inch (53 cm) diameter 1 500 rpm straight line electric high speed buffing machine, fitted with a cream-colored synthetic high speed burnishing pad[11].

After μ_s determinations were made, the tiles were returned to the wear test and the auto-scrubbing and high speed burnishing was continued. The machine used for this stage of testing was a Hako 21 inch (53 cm) diameter 2 100 rpm straight line electric high speed buffing machine, fitted with a cream-colored synthetic high speed burnishing pad[11].

After μ_s determinations were made, the tiles were returned to the wear test and the auto-scrubbing and high speed burnishing was continued. The machine used for this stage of testing was a Pioneer/Eclipse 22 inch (56 cm) diameter 2 200rpm straight line propane high speed buffing machine, fitted with a cream-colored synthetic high speed burnishing pad[11].

After μ_s determinations were made, the tiles were returned to the wear test and the auto-scrubbing and high speed burnishing was continued. The machine used for this stage of testing was a Pioneer/Eclipse 22 inch diameter 2200 rpm straight line propane high speed buffing machine, fitted with a cream-colored synthetic/natural fiber blend high speed burnishing pad[11]. (This pad is more aggressive than the synthetic high speed burnishing pad.)

Results and Conclusions

The results of this series of tests is presented in Table 13. The tests are listed in order of increasing abrasion (based on pad pressure and the aggressiveness of the pad) rather than the order in which the data was generated.

High speed burnishing maintenance had no effect on μ_s of any of the polishes tested.

The data shows a statistically insignificant, but suggestive trend toward higher values of μ_s as the aggressiveness of the procedure is increased (more of the film surface is removed).

Subjective "paper under foot" slip resistance testing [6] immediately after high speed burnishing indicated that the burnished polishes, particularly after the more aggressive burnishing operations, would have a higher μ_s than was reflected in the data. Thus alerted, we undertook tests to determine if high speed burnishing, like spray buffing, shows a time-dependent change in μ_s on completion of the maintenance procedure.

Table13–μ_s *of polishes after high speed burnishing*[7].

Polish	HS-04-9	DP-2-4	Commercial E
Baseline μ_s			
Archive	0.63 +/- 0.02	0.59 +/- 0.03	0.56 +/-0.03
2100 rpm electric, synthetic pad			
20 operations[a]	0.63 +/- 0.02	0.59 +/- 0.03	0.55 +/-0.3
1500 rpm electric, synthetic pad			
20 operations[a]	0.63 +/- 0.03	0.59 +/- 0.03	0.56 +/-0.03
1070 rpm electric, synthetic pad			
20 operations[a]	0.65 +/- 0.03	0.59 +/- 0.03	0.58 +/-0.03
2200 rpm propane, synthetic pad			
20 operations[a]	0.65 +/- 0.03	0.62 +/- 0.03	0.58 +/-0.03
2200 rpm propane, synthetic / natural fiber blend pad			
20 operations[a]	0.66 +/- 0.03	0.63 +/- 0.04	0.59 +/-0.03

[a] Data is the average of μ_S determinations from 24 tiles.

Equilibration of μ_s after High Speed Burnish Maintenance

The increase in μ_s after high speed burnishing appears to be more pronounced for DP-2-4 films than for the other polishes, and it also appears to be a function of the aggressiveness of the pad/machine combination.

Of the three polishes studied above, the order of abrasion resistance is DP-2-4 > HS-04-9 > "Commercial E". This order is based on the relative amounts of material abrasively removed by a given pad/machine combination, measured by determining the number of high speed burnishing passes required to abrade through one coat of (equal solids) polish.

The magnitude of change in μ_s observed after spray buffing appeared to be related to the relative hydrophilicity of the polishes. Of the three polishes studied for high speed burning response, the order of hydrophilicity (based on polish surface energy measurements) is HS-04-9 > DP-2-4 > "Commercial E".

High speed burnishing involves only the chemicals in the polish composition. Formulation details are not known for "Commercial E", so its inclusion in this study of change in μ_s after high speed burnishing would not be useful in identifying any chemical causes of the μ_s change. On completion of this study, "Commercial E" was included as a test of the hypothesis that we developed from the results on the other polishes.

The procedure for this study is almost identical to that of the study of μ_s equilibration after spray buffing. The 2200 rpm propane burnisher with a blended fiber pad was used, since this appeared to show the largest change in μ_s in screening studies.

Results and Conclusions

The results of this testing are shown graphically in Fig. 6, with pertinent data from he plots presented in Table 14.

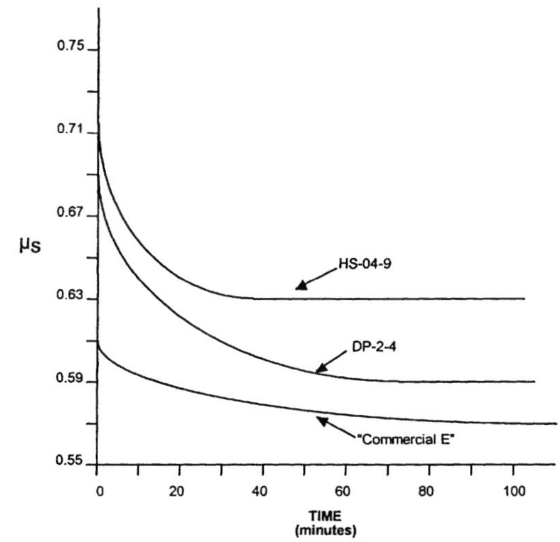

Fig. 6-*Time-dependent change of* μ_s *after high speed burnishing.*

Table 14-*Data from Fig. 6.*

Polish	HS-04-9	DP-2-4	Commercial E
Baseline μ_s			
Archive	0.63 +/- 0.02	0.59 +/- 0.03	0.56 +/-0.03
Extrapolated μ_s at time = 0	0.71	0.69	0.61
Deviation from archive μ_s	0.08	0.10	0.05
Percent deviation	13	17	9
Time to restore μ_s (minutes)	35	75	120

The data plots describe zero order reactions or processes (determined by least squares fitting of the raw data to different order rate laws, though the data for HS-04-9 could also be fit to a first order rate law). If we consider the archive value of μ_s to be an equilibrated value, then the deviations of μ_s from the equilbrium value are due to a chemical or physical process that follows zero order kinetic rate law. This was also true of the deviations of μ_s seen in spray buffing maintenance (above), but the process was operating in reverse.

High speed burnishing is known [5] to provide heat to the polish film as a result of the abrasive process and friction between the polish and pad. (When high speed burnishing was first invented it was thought that this heat was the cause of the polish response, rather than a consequence of the abrasive process.) The loss of heat through convection or radiation to the environment follows a zero order rate law[12], so the deviation to higher values of μ_s could be due the burnished films being warm. Temperature effects on static coefficient of friction are not well understood because they have generally been ignored.

It is possible that warm polish films have higher μ_s than cool polish films, but the differences in the amount of time required for the polishes to return to equilibrium temperature are impossible to explain. Polish films have a higher thermal conductivity than does the vinyl composition tile that it is on, and the polish on a tile is a minuscule proportion of the weight of the polished tile. From the known differences in polish composition between HS-04-9 and DP-2-4 there is nothing that would cause one to take twice as long to cool as the other.

[12] The rate law is actually first order in the temperature differential, but for small differentials this is experimentally indistinguishable from a zero order process.

The film heating and cooling hypothesis also fails to enlighten us about the cause of the μ_s deviations after spray buffing. Spray buffing is also an abrasive process [4] that generates frictional heat, but the deviations are in the opposite direction.

In the examination of the spray buffing deviations the inherent hydrophilicities of the polish films, and the changes in hydrophilicity induced by the chemistry of the spray buffing media were thought to cause films to lose water at different rates. The lower μ_s values were thought to be due to water-saturated or hydrated polish surfaces.

If we apply that same logic to the deviations in μ_s caused by high speed burnishing, the deviations are seen to be due to the loss of the hydrated polish surface. The function of the high speed burnishing procedure is to remove the polish film surface, because that is the mechanism for its repair of traffic damage and gloss restoration.

The time-dependent decrease in μ_s after high speed burnishing is due to the restoration of hydration in the (newly exposed) polish surface by adsorption of water from the atmosphere. Because the concentration of water in the atmosphere is low, this process is much slower than the loss of water to the atmosphere in the drying of the over-hydrated spray buffed polish surface[13].

It has been previously recognized that polish surfaces are lightly hydrated and in equilibrium with the moisture in the atmosphere. This study has shown that perturbation of this hydration leads to changes in μ_s.

It would be informative to know the μ_s of polish films when dehydrated, or after equilibration at 0 percent relative humidity, since this would represent the maximum μ_s of the polish. This is not possible with ASTM D2047, because the specification leather must be equilibrated at ~50 percent relative humidity to retain a reproducible degree of pliability. It is known[14] that when equilibrated at lower relative humidity the specification leather gives μ_s data that is *lower* and much more variable (larger standard deviation). Most operators keep the James machine, or at least the leather, in a constant temperature/humidity environment to avoid these problems. The reduction in μ_s due to leather dehydration is in the opposite direction of the increase in μ_s anticipated from the dehydration of the film surface. Rather than specify that the leather is to be equilibrated in a narrow temperature/relative humidity (indirectly specifying the partial pressure of water vapor in the air) environment, both the leather and the polish films to be tested should be equilibrated under conditions consistent with the use-environment of the polish.

The conventional procedure of ASTM D2047 begins with polishes surfaces that have not been modified. They are applied to the tile per ASTM D4103, and allowed to dry for testing. This means that the film surface is equilibrating by losing hydration, and we have seen that this is a fast process.

The variability of μ_s as a function of film surface hydration noted in this study is relatively small, and of no significance in the overall slip resistance of the polish in traffic. The slow restoration of hydration after high speed burnishing is not problematic,

---

[13] The rate law for hydration by atmospheric moisture is first order in the partial pressure of water, but the experiment provides an effectively 'infinite' reservoir of fixed humidity air so that the rate law resolves to being zero order.

[14] Unpublished CSPA Polishes and Floor Maintenance Division ruggedness test of the procedure of ASTM D 2047, completed in 1979.

because until the hydration is complete, the polish has a higher than normal (equilibrated) μ_s.

Summary

The standard floor maintenance procedures of damp mopping, mop scrubbing, machine scrubbing, and auto-scrubbing have no measurable effect on the static coefficient of friction of polishes, and do not change polish slip resistance.

Traffic on polishes, when not accompanied by cleaning maintenance, lowers the static coefficient of friction of polishes, and reduces the slip resistance of the film. Ingrained soil in the film is the cause of this deterioration in slip resistance. Scuffing, loss of gloss (micro-scratches), and black marking have no measurable effect.

The abrasive machine maintenance procedures of spray buffing and high speed burnishing temporarily change the static coefficient of friction of polishes by removing a surface layer of hydrated polish. This perturbation by spray buffing relaxes very rapidly, while that by high speed burnishing is protracted and is probably a function of the partial vapor pressure of water in the atmosphere (temperature and relative humidity). Spray buffing temporarily reduces polish μ_s, and high speed burnishing temporarily increases polish μ_s. If the film surface hydration/dehydration hypothesis for these deviations of μ_s for spray buffed and high speed burnished polishes is correct, then the maximum amount of change possible in μ_s is a function of the inherent hydrophilicity of the polish film. Polish films are independently restricted to a narrow range of hydrophilicity that is compatible with other polish performance and wet test properties, so that broader deviations than those seen in this study are highly improbable.

Acknowledgements

The work in this paper was done intermittently over a period of more than four years. The majority of this often-tedious work was done almost willingly by an exceptional floor care products chemical technician, Charles E. O'Farrell, Jr.

This research received the full support of Rohm and Haas research management, in the persons of Richard E. Zdanowski, Manager, Chemical Specialties Research; Dr. Joseph A. Lavelle, Manager, Construction Products Research, and Dr. William D. Emmons, Research Director, Polymers, Resins, and Monomers. They actively supported this work as a service to the floor maintenance industry, in spite of it promising no possible commercial benefit for Rohm and Haas Company.

References

[1] Sacher, A., "Slip resistance and the James machine 0.5 static coefficient of friction", *Standardization News,* August 1993, pp. 2-9, and reference cited therein.

[2] ___________, *Population Facts and Gazateer*, United States Census Bureau, 1993.

[3] Flick, E. W., *Advanced Cleaning Product Formulations,* Vol. 5, Noyes Publications, Norwich, NY, 1989, pp.148-160.

[4] Falconio, R. R., Owens, J. M., and Sava, G. A., "New Trends in Floor Care: Spray Buffing and Burnishing", *Resin Review*, Vol. 33, No. 2 (1983).

[5] Owens, J. M., "The Physical Processes of High Speed Burnishing", *Chemical Times and Trends,* **11**, 44 (1988).

[6] CSPA Polishes and Floor Maintenance Division, "CSPA Test Method 245-70, Comparative Determination of Slip Resistance of Floor Polishes," *Test Methods Compendium*, Consumer Specialties Products Association (2002).

Appendix

The formulations used in this work are largely based on recommended starting point formulations using floor care polymers from Rohm and Haas Company. This selection is not an endorsement of these products, but simply an acknowledgement that we are intimately familiar with these products and their chemistry. This chemical knowledge allows us to understand effects and consequences and make reasonable hypotheses, and it is not available for products from other manufacturers or from proprietary commercial formulations.

In the discussions in this paper concerning proprietary commercial formulations, the information presented was from label copy, discussions with the manufacturer, or in-house analytical work. The analysis of commercial formulations should be considered to provide, at best, only a reasonable approximation.

Polish Formulations

A. Formulation N-32-25

Material	Solids (%)	Parts (wt)	Function
Water	0	32.03	solids and viscosity control
SWS-211[15]	10	0.02	defoamer
Acrysol®[16] 644	42	5.56	alkali soluble resin
Fluorad®[17] FC-120	1	0.70	wetting agent
Diethylene glycol ethyl ether	0	6.35	coalescing solvent
Dipropylene glycol methyl ether	0	0.56	coalescing solvent
Tris-butoxyethyl phosphate	100	1.56	leveling agent, plasticizer
Kathon®[13] CG/ICP[13]	1.5	0.03	biocide
Rhoplex®[13] B-832	40	43.79	polymer
AC®[18]-392N	35	5.01	synthetic wax
Epolene®[19] E-43N	40	4.39	synthetic wax

[15] Stauffer-Wacker Silicones, Adrian, MI

[16] Rohm and Haas Company, Philadelphia, PA.

[17] 3M Company, St. Paul, MN.

[18] Allied-Signal Corp., Morristown, NJ.

[19] Eastman Chemicals, Kingsport, TN.

Totals	24.94	100	minimum maintenance polish

Polymer / ASR / Wax solids ratio: 75 / 10 / 15

B. Formulation N-32-2

Material	Solids (%)	Parts (wt)	Function
Water	0	59.10	solids and viscosity control
SWS-211[15]	10	0.02	defoamer
Acrysol®[16] 644	42	3.34	alkali soluble resin
Fluorad®[17] FC-120	1	0.42	wetting agent
Diethylene glycol ethyl ether	0	3.82	coalescing solvent
Dipropylene glycol methyl ether	0	0.34	coalescing solvent
Tris-butoxyethyl phosphate	100	0.94	leveling agent, plasticizer
Kathon®[16] CG/ICP	1.5	0.03	biocide
Rhoplex®[16] B-832	40	26.34	polymer
AC®[18]-392N	35	3.01	synthetic wax
Epolene®[19] E-43N	40	2.64	synthetic wax
Totals	15.00	100	minimum maintenance polish

Polymer / ASR / Wax solids ratio: 75 / 10 / 15

C. Formulation DP-2-4

Material	Solids (%)	Parts (wt)	Function
Water	0	29.02	solids and viscosity control
SWS-211[15]	10	0.02	defoamer
Fluorad®[17] FC-120	1	0.90	wetting agent
Diethylene glycol ethyl ether	0	4.75	coalescing solvent
Dipropylene glycol methyl ether	0	2.23	coalescing solvent
Dibutyl phthalate	100	1.07	plasticizing solvent
Tris-butoxyethyl phosphate	100	1.43	leveling agent, plasticizer
Kathon®[16] CG/ICP	1.5	0.03	biocide
Duraplus®[16] 2	38	50.07	polymer
Resinall®[20] 802	25	4.48	alkali soluble resin
AC®[18]-392N	35	3.22	synthetic wax
Epolene®[19] E-43N	40	2.78	synthetic wax
Totals	15.00	100	minimum maintenance polish

Polymer / ASR / Wax solids ratio: 85 / 5 / 10

D. Formulation HS-04-9

[20] Resinall Corporation, Severn, NC.

Material	Solids (%	Parts (wt)	Function
Water	0	43.35	solids and viscosity control
Fluorad®[17] FC-120	1	0.45	wetting agent
SWS-211[15]	10	0.02	defoamer
Diethylene glycol ethyl ether	0	1.67	coalescing solvent
Dipropylene glycol methyl ether	0	1.67	coalescing solvent
Dibutyl phthalate	100	0.76	plasticizing solvent
Tris-butoxyethyl phosphate	100	0.76	leveling agent, plasticizer
Kathon®[16] CG/ICP	1.5	0.03	biocide
Rhoplex®[16] B-1604	38	37.36	polymer
Resinall®[20] 802	25	3.79	alkali soluble resin
AC®[18]-392N	35	5.41	synthetic wax
Epolene®[19] E-43N	40	4.73	synthetic wax
Totals	20.40	100	minimum maintenance polish
Polymer / ASR / Wax solids ratio:	75 / 5 / 20		

Spray Buffing Media Formulations

E. Formulation SBM-44-2

Material	Solids (%)	Parts (wt)	Function
Water	0	91.81	solids and viscosity control
Fluorad®[17] FC-120	1	0.10	wetting agent
SWS-211[15]	10	0.03	defoamer
Acrysol®[16] 644	42	2.98	ASR, lubricant
Abex®[21] 18s	100	0.25	stabilizer, lubricant
Diethylene glycol methyl ether	0	0.33	solvent
Ethylene glycol	0	0.20	stabilizer
AC®[18]-540N	30	4.17	lubricant, opacifier
Formalin	37	0.15	Biocide
Ammonia	0	to pH 8 – 8.5	neutralize ASR, lubricant
Totals	2.75	100	high speed spray buff medium

F. Formulation SBM-44-1

Material	Solids (%)	Parts (wt)	Function
Water	0	85.05	solids and viscosity control
Acrysol®[16] 644	42	14.20	ASR, lubricant
Mono ethanol amine	0	0.60	neutralize ASR, lubricant
Formalin	37	0.15	Biocide

---

[21] Alcolac, Inc., Linthicum, MD

Totals	6.02	100	low speed spray buff medium

Floor Cleaner Formulation

G. Formulation NIO-0.5

Material	Solids (%)	Parts (wt)	Function
Water	0	99.42	solids and cost control
OPE_{40} surfactant[22]	70	0.26	high HLB surfactant
NPE_{10} surfactant[23]	100	0.16	intermediate HLB surfactant
OPE_5 surfactant[24]	100	0.16	low HLB surfactant

OPE_{40} surfactant is an octyl phenol that had been reacted with 40 moles of ethylene oxide. The commercially available surfactant is a blend of from 38 to 45 mole adducts of ethylene glycol on the alkyl phenol. This surfactant was selected because it is particularly adept at dispersing light oils, such as sebum, and moderately hydrophilic soils.

NPE_{10} surfactant is nonyl phenol that had been reacted with ten moles of ethylene oxide. The commercially available Nonyl phenol is a blend of 9, 10, and 11 mole ethylene oxide adducts to the alkyl phenol. This surfactant was selected because it is the most widely used nonionic surfactant class in the world and is at least a constituent of most commercial floor cleaner formulations. It serves not only as a surfactant to act on oily soils, but it also helps to solubilize or disperse low HLB surfactants.

OPE_5 surfactant is an octyl phenol that had been reacted with five moles of ethylene oxide. The commercially available octyl phenol surfactant is a blend of four and five mole adducts of ethylene glycol on the alkyl phenol. This surfactant was selected because it is commonly used in degreasing or heavy duty cleaning formulations for its ability to make stable dispersions or emulsions of very hydrophobic oils, fats, and greases.

---

[22] Triton® X-405 from Union Carbide Corp., Industrial Chemicals Division, Danbury, CT.

[23] Triton® N-101 from Union Carbide Corp., Industrial Chemicals Division, Danbury, CT.

[24] Triton® X-45 from Union Carbide Corp., Industrial Chemicals Division, Danbury, CT.

Paul F. Lewis[1]

Portable Slip Testers

Reference: Lewis, Paul F., "**Portable Slip Testers,**" *Technology of Floor Maintenance & Current Trends, ASTM STP 1448,* W. J. Schalitz, Ed., ASTM International, West Conshohocken, PA, 2004.

Abstract: The world has seen a vast number of portable floor slip-resistance testers over the years. But what do they actually measure? How do they work? Which testers can be used interchangeably? And which work as effectively as the United States-recognized James Machine? This paper addresses all of these issues and discusses latest trends in floor safety.

Keywords: Portable, Slip, Floor, Slip-resistance, Coefficient of Friction, James Machine, Pendulum, Tortus, Ramp, Sellmaier, Brungraber, English, Surtronic, Surface Roughness, Dynamometer

Introduction

There are a number of commercially available slip testers available today, touted to measure a floor's slip resistance. This paper reviews these machines and compares how they perform relative to the nationally recognized James Machine.

What is Coefficient of Friction?

Coefficient of Friction (CoF) can be defined in two ways: Static CoF, which is defined as the ratio of horizontal to vertical forces required to start a body moving; Dynamic (or Kinetic) CoF, which is defined as the ratio of horizontal to vertical forces required to keep a body moving.

[1]Group Leader/Research Associate, Global Floor Care Research & Development, Johnson Diversey Inc., 8310 16th Street, Sturtevant, WI 53177, USA.

FIG. 1 below illustrates this and gives some examples [1]. Typically, all things being equal, Dynamic CoF is less than Static CoF.

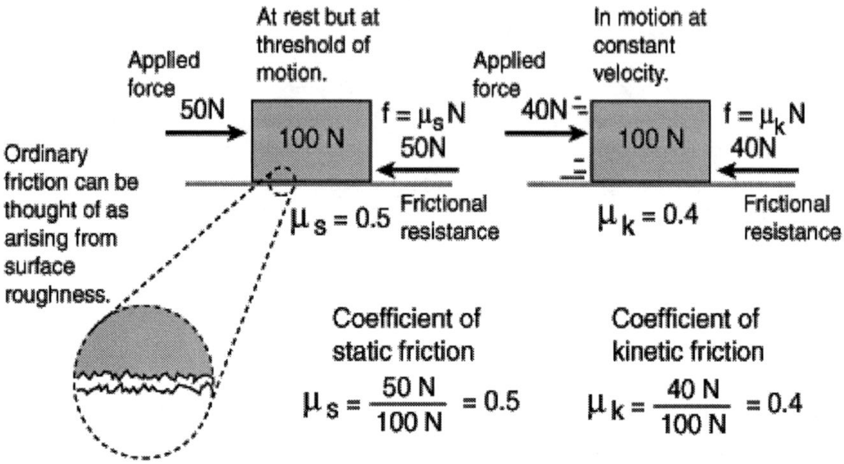

FIG. 1-Measuring Static and Dynamic Coefficient of Friction

What is Slip Resistance?

ASTM defines slip resistance (SR) as that property of a floor surface that is designed to prevent slipping (Static Coefficient of Friction of Polish-Coated Floor Surfaces as Measured by the James Machine, ASTM D 2047-99). Floor surfaces having a static CoF of 0.5 or greater as measured in accordance with ASTM D 2047 are considered to be slip resistant.

The James Machine

The value and significance of the James Machine (FIG. 2) is legendary. The articulated strut tester incorporates an 80 lbs. (36.3 kg) weight that is applied through a pivoted strut to a clean leather shoe (or sensor) placed on a panel to which a floor finish or other floor coating has been applied. The panel and the shoe move together horizontally, powered by the motor and the weight develops a horizontal force at the shoe/polish interface, which increases laterally until a slip occurs. The distance the panel moves before slippage occurs is accurately measured and recorded on a chart that is calibrated to read static coefficient of friction based on the linear distance the panel moves prior to slippage.

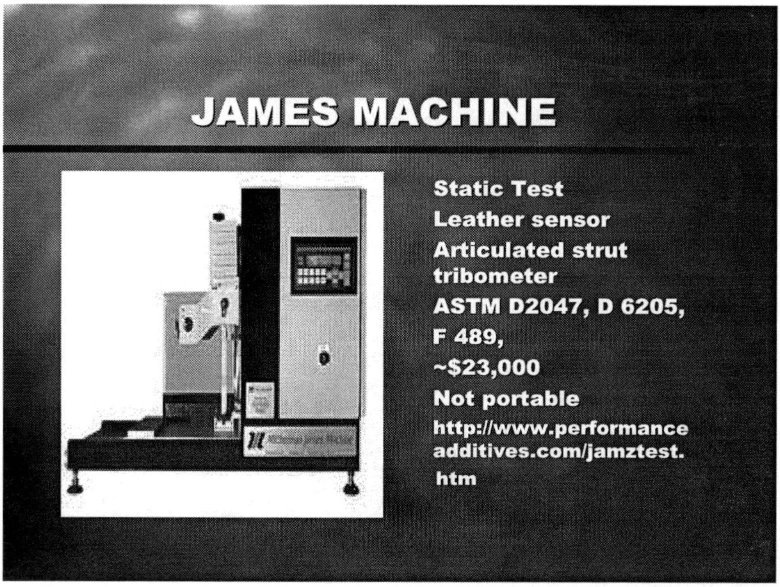

FIG. 2-The James Machine

The James Machine method was approved by the American Society for Testing and Materials (ASTM) as the basis for Standard Test Method D-2047. The static CoF of 0.5 has widely been recognized as the minimum value acceptable in receiving the UL seal of approval, even though data exists to show that people do not slip if they walk on the floor with a static CoF as low as 0.3 – 0.45. Given the safety margin that the 0.5 value provides, the Consumer Specialty Products Association (CSPA) adopted the 0.5 figure as the point which determines whether a product should be classified as slip resistant. The James Machine, used for dry, uncontaminated surfaces, is currently the only official ASTM method for measuring static coefficient of friction of polished floors.

Other Slip Meters

A number of other slip-resistance testers have been developed over the years and have been used on floors. Some of these are portable, some are not. Some measure Static CoF, some measure Dynamic CoF, some measure Slip Index, Slip Resistance or British Pendulum Number (BPN) values. But a review of these machines shows that none of them can be used interchangeably with the James Machine. In fact, none of these machines can be used interchangeably with any other slip tester; they will all give different results, primarily because of the different mechanisms they employ and different sensors they use.

These portable slip testers are marketed to facility owners/managers with the intent that they be used to ensure the safety of their floors. Because none of them correlates with the James Machine & because there is no recognized standard for any of them, they cannot be used for this purpose and their value is actually very limited.

FIGs. 3 - 12 show these other testers.

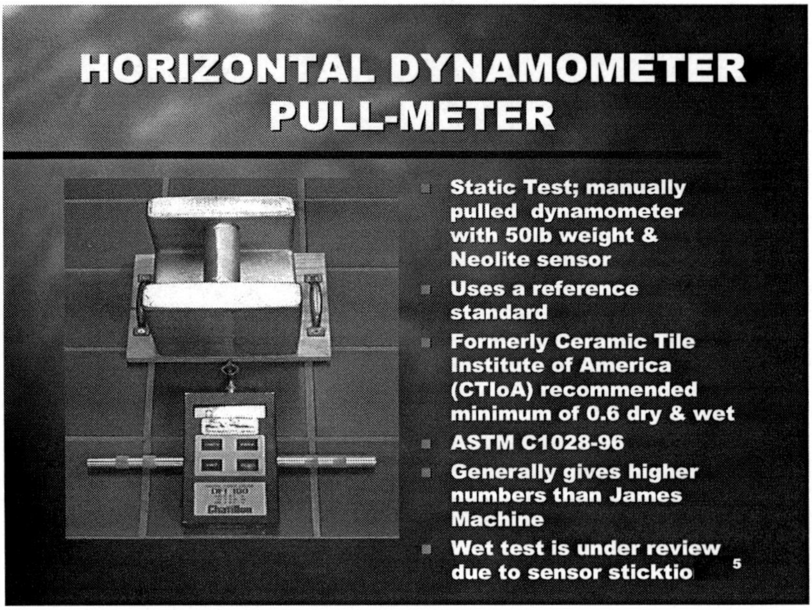

FIG. 3-The Horizontal Dynamometer Pull-Meter

As can be seen from the review, none of these machines operates identically to any other machine. This helps to explain the widely differing results obtained from one to another. The only machine that operates similarly to the James Machine is the Brungraber Mk I. Even this, however, has been shown to give different results. This is partly caused by the differences in the weight used on the machines (10 lbs. [4.5 kg] on the Brungraber, vs 80 lbs. [36.3 kg] on the James) and partly by the differences in mechanisms used to propel the sensor. The James machine accurately measures Static CoF directly onto the chart. In the case of the Brungraber Mk I CoF has to be calculated separately from the PAST number.

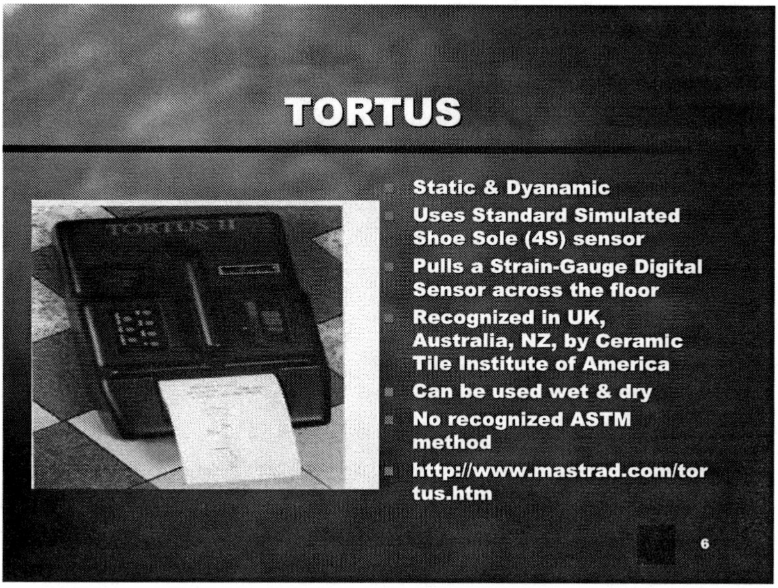

FIG. 4-The Tortus Slip Tester

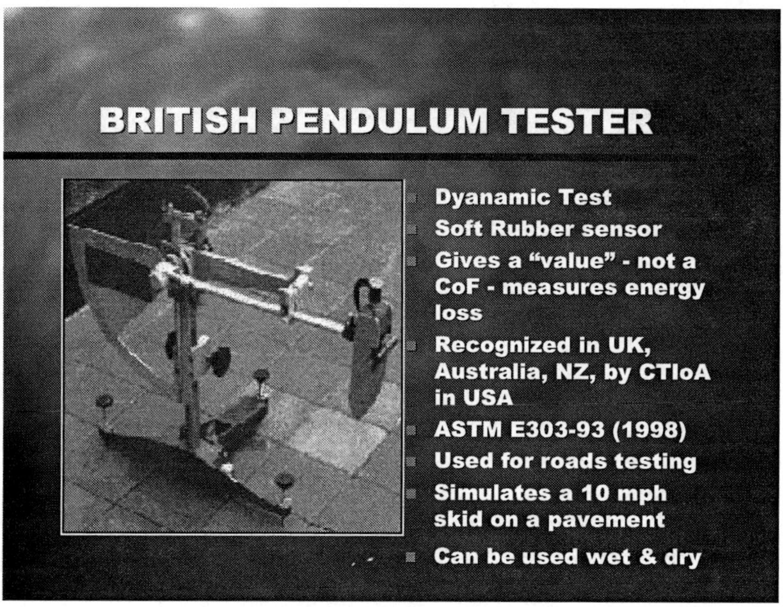

FIG. 5-The British Pendulum Tester

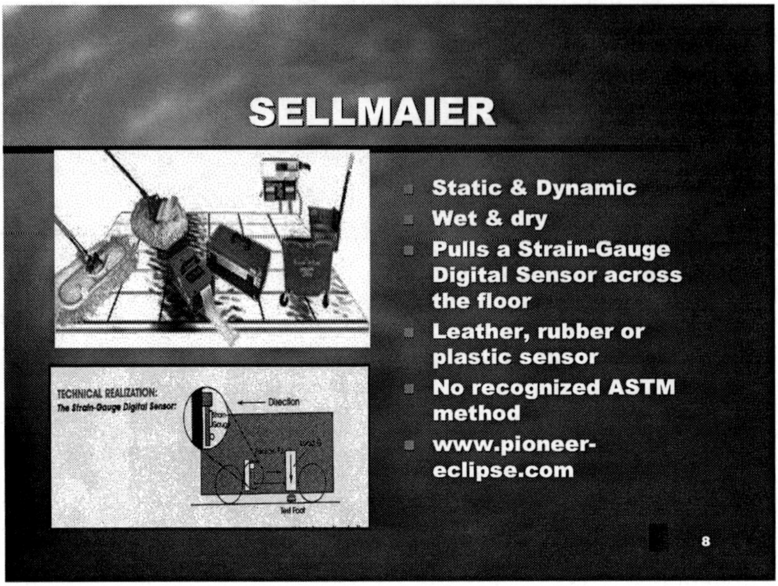

FIG. 6-The Sellmaier Floor Slip Tester

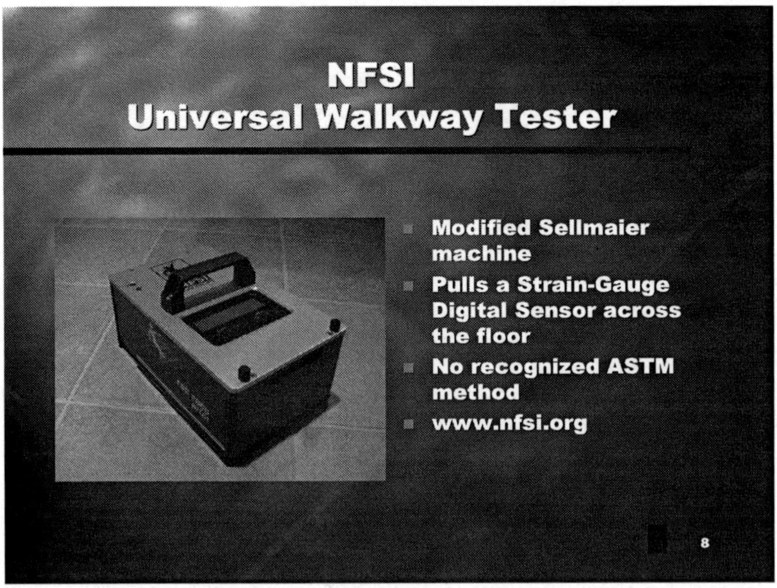

FIG. 7-The NFSI Universal Walkway Tester

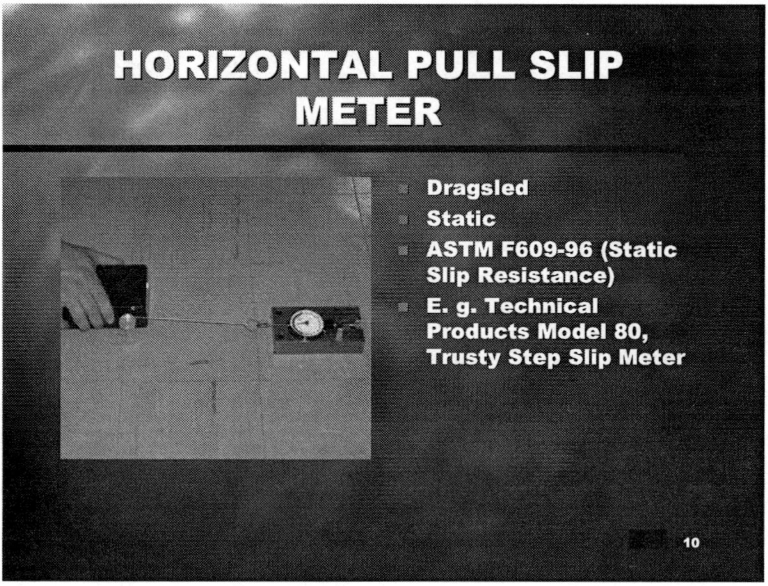

FIG. 8-A Horizontal Pull Slipmeter

FIG. 9-The Topaka Windlass Slip Tester

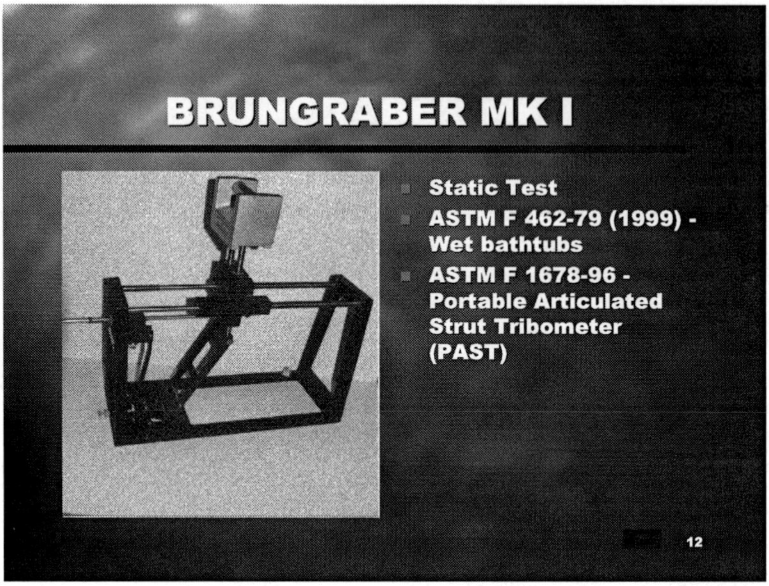

FIG. 10-The Brungraber Mk I Portable Articulated Strut Tribometer

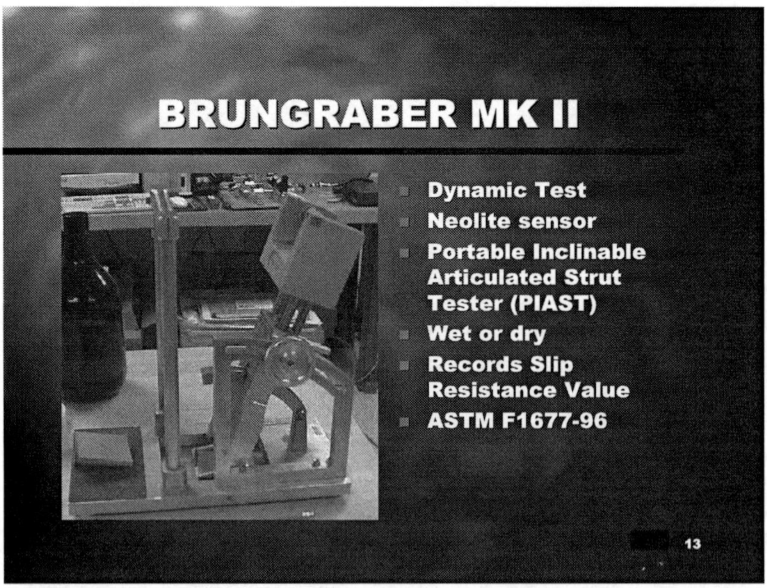

FIG. 11-The Brungraber Mk II Portable Inclinable Articulated Strut Tribometer

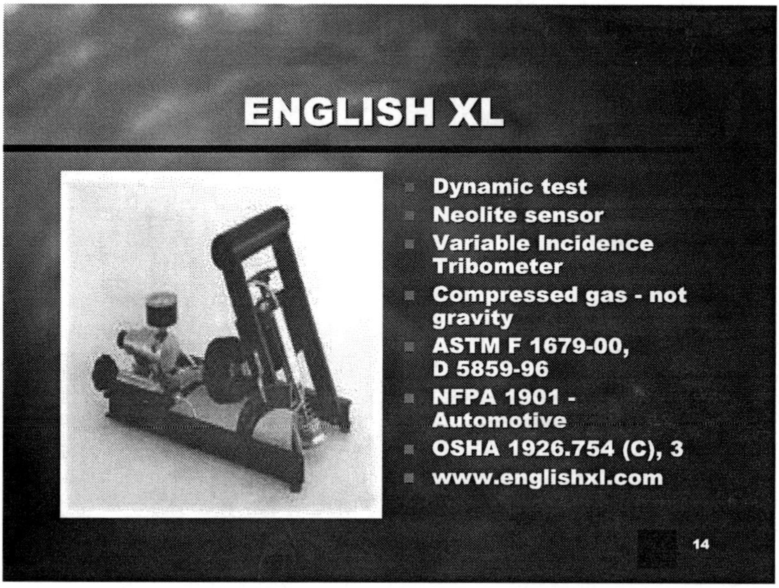

FIG. 12-The English XL Variable Incidence Tribometer

The Non-ASTM Perspective

ASTM has a number of groups/committees investigating important slip resistance matters eg D21, F13 Committees. In addition to these there are a number of groups actively studying this topic, including UK Slip Resistance Group (RAPRA), Ceramic Tile Institute of America (CTIoA), Standards Australia and Standards New Zealand. Below is a review of additional test equipment that has been identified by these groups.

Surface Roughness

Regardless of CoF numbers that may be obtained from slip testers, smooth floors tend to be slippery when wet. An effective way to improve the wet slip resistance of a floor is to increase its surface roughness. There is a simple way to measure this in the field, as shown in FIG. 13.

Both the UK Slip Resistance Group [2] and Bunterngchit [3] have demonstrated the correlation between surface roughness and the tendency for a wet slip to occur.

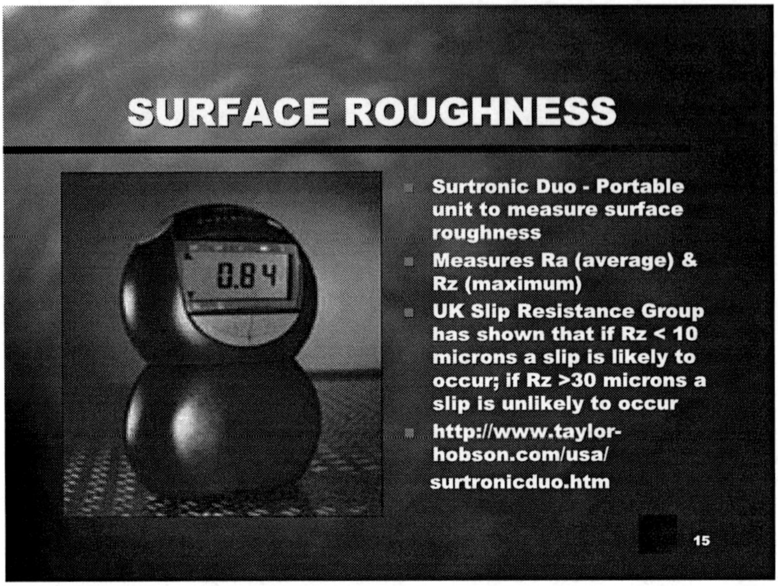

FIG. 13-The Surtronic Duo Surface Roughness Meter

Inclined Ramp Testing

Inclined ramp testing, originally pioneered in Germany, is currently used in some European countries, Australia & New Zealand to assign wet slip resistance values to flooring types, based on actual pedestrians walking on them. R values of R9 – R13 are assigned, based on a test subject walking on an inclinable ramp with a section of flooring on the ramp (shown in FIG. 14).

Based on the angle at which a harnessed man can walk on the surface, covered in motor oil, using standardized footwear, the R value can be calculated:

Angle	R Value
3 – 10 degrees	R9
10 - 19 degrees	R10
19 - 27 degrees	R11
27 - 35 degrees	R12
> 35 degrees	R13

Specific R values have been recommended for specific areas. For example counter areas in financial institutions need to be R9, while fish processing areas need to have floors at R13. Other categories exist for barefoot areas [4].

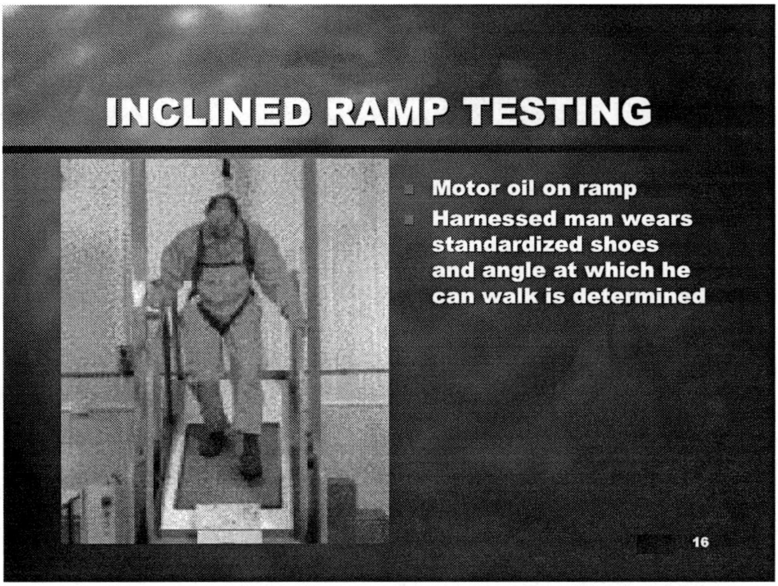

FIG. 14-The Inclined Ramp Tester

The Ceramic Tile Institute of America has endorsed the use of these new categories for new flooring in an attempt to reduce wet slips. For portable testing of floors that could get wet, the CTIoA also now recommends floors be tested with British Tortus & Pendulum, with minimum recommended values of 0.5 and 35 BPN respectively [5].

Summary

In conclusion, a study of numerous existing methods of measuring slip resistance indicates the following

1. No equipment currently exists that gives the same values as the all-important James Machine.
2. Slip testers all give different values. This is partly because they all operate in different ways, at different speeds, using different sensors and are either powered by gravity, compressed air or some other mechanism.
3. In order to get an accurate picture of a floor's complete slip resistance profile, a combination of factors should be evaluated, for example: its Static CoF, Dynamic CoF, Surface Roughness, the nature of any contaminants and the use to which that floor is routinely being put.

4. Because smooth floors can be slippery when wet, further definition is needed by committees to identify the types of floors needed for areas known to be wet or otherwise contaminated. Work done by CTIoA, defining R values of flooring, forms a good basis for this. Its broader adoption, possibly with minor modifications, should be considered.

References

[1] Jones, E. R. (Rudy) and Childers, R. L, "Contemporary College Physics", 2nd Ed., Addison-Wesley, 1993, Sec. 4.8.

http://hyperphysics.phy-astr.gsu.edu/hbase/frict.html#rou

[2] The UK Slip Resistance Group, "Measurement of Floor Slip Resistance", Issue 2, June 2000.

[3] Bunterngchit, Y., "The Influence of Floor Surface Roughness on the Slip Resistance Coefficient", XXII International Congress on Occupational Health, Sydney, Australia, King Mongkut's Ins. of Tech., North Bangkok, Thailand, 1987 .

http://library.kmitnb.ac.th/article/atc34/atc00136.html

[4] Bowman, R., "Specifiying Slip Resistant Tiling".

http://www.dbce.csiro.au/brochures/slip/

[5] La Fortune, G. & Sotter, G., "Tile Institute Endorses Slip Resistance Tests, Safety Standards", Ceramic Tile Institute of America Publication, January 8th 2002.

LEGAL AND REGULATORY ISSUES

William C. Balek[1]

Managing Slips and Falls: A Legal Perspective

Reference: Balek, W. C., **"Managing Slips and Falls: A Legal Perspective,"** *Technology of Floor Maintenance and Current Trends, ASTM STP 1448,* W. J. Schalitz, Ed., ASTM International, West Conshohocken, PA, 2004.

Abstract: Annually, slip and fall incidents exact a substantial toll in terms of death, personal injury, personal suffering, workers compensation, loss in productivity and civil liability. The cause of slip and fall incidents and resultant damages are multiple and include unclean/unsafe floors, physiological characteristics of individuals, lack of training, inappropriate footwear and even fraud. A clear understanding of the underlying causes of slip and falls provides us with the framework for identifying those factors we can control and thus minimize and otherwise reduce one's potential liability.

Numerous parties may be responsible and therefore potentially liable for a slip and fall incident depending on the particular circumstances of a given situation. Parties that may be liable in a slip and fall incident include owners or operators of retail and other commercial facilities, and distributors and manufacturers of floor care products, flooring and footwear.

This paper outlines various proactive measures that individuals and companies can implement and which will help to eliminate or substantially reduce the incidence of slip and fall injuries, and any potential liability.

Keywords: Slip and fall; Liability

Introduction

Overview

This presentation will review those practices and procedures that companies can implement for the purpose of reducing their potential slip and fall liability as it relates to falls on same level surfaces. We will first define the problem by reviewing the various burdens imposed by slip and fall incidents as well as the major causes of such events. Legal theories of liability will then be briefly explored followed by prophylactic measures companies can employ to prevent such incidents from occurring in the first place.

Statement of Problem

Slip and fall incidents continue to impose substantial costs each year upon employers, owners/operators of commercial establishments, individuals, and society in general. The following statistics reflect the gravity of the situation:

---

[1] Director of Legislative Affairs, International Sanitary Supply Association, 7373 N. Lincoln Ave., Lincolnwood, IL 60712.

Employers

- Liberty Mutual has concluded that falls are the second leading cause of disabling workplace injuries resulting in employees missing five or more days of work in 1999, resulting in 11.5% of total direct costs associated with workplace injuries or $4.6 billion. Indirect costs associated with non-fatal workplace falls are estimated to be $9.2 to $23 billion [1].

- According to the Bureau of Labor Statistics, falls accounted for 12.4% of all workplace fatalities (733 work related deaths)[2].

- According to the 2002 Liberty Mutual Workplace Safety Index, amongst employers there exists a significant gap between perception and reality resulting in a misallocation of resources. For example employers reported "falls on the same level" as the seventh most important cause of workplace injuries, while it was actually ranked as the second most costly injury cause. This survey data indicates employers may tend to place less priority on injury causes that actually have a greater potential impact on workplace safety.

- It is estimated that slip and fall incidents account for almost 30% of all workers compensation claims [3].

- Slips and falls are the leading cause of employee injury in the foodservice industry [4].

General

- Falls were the second leading cause of deaths that occurred at work, home, commercial and other locales in 1998, resulting in 16,274 deaths[5].

- Falls took the lives of 9,300 people at home, the majority of them over the age of 65[6].

- The total direct cost of all fall injuries for people age 65 and older in 1994 was $20.2 billion. By 2020, the cost of fall injuries is expected to reach $32.4 billion [7].

- Falls account for more than eight million hospital emergency room visits, representing the leading cause of visits (21.3%). Slips and falls account for more than one million visits, or 12% of total falls [8].

Major Causes of Slip and Fall Incidents

There are numerous causes of slip and fall incidents that occur on same level surfaces, some of which are controllable and others that are not. While there is some disagreement about how much a particular cause contributes to the problem, most experts agree that the major causes of slip and fall incidents are as follows:

- Unsafe, Unclean Floors,
- Inappropriate Footwear,
- Fraud,
- Inadequate Identification of a Hazard,
- Insufficient Training,
- Physiological Characteristics (age, disability, etc.), and
- Inappropriate, Non-Compliant Floor Care Product.

Certain factors, such as the physiological characteristics of the "walker," are not within our control. However, many of the causes are factors that can be controlled, such as unclean floors, appropriate floor care products, and training. In managing or minimizing potential liability, we must focus on instituting a comprehensive approach that addresses those factors that we can manipulate and otherwise control. But before we discuss prophylactic measures, it is helpful to have a general understanding of the basis of civil liability.

General Discussion of Liability

When a slip and fall incident occurs, determining who is responsible can be a complicated matter. In the first place, multiple parties are involved in floor safety: owner/operator of the establishment, employer, manufacturer/supplier of floor care products, flooring or footwear, and the injured party. Furthermore, responsibility and legal liability depend upon the unique circumstances of each situation. However, a brief discussion of the basis of civil liability under our legal system is helpful in understanding where responsibility lies and what measures we can take to minimize liability.

Please be advised that the following discussion is based on generalizations. You should be aware that the legal basis of liability varies from state to state.

Potentially Liable Parties in a Slip and Fall Incident

- Owner/Operator of a Facility
- Contract Cleaning Service Company
- Employer
- Supplier/Manufacturer of Floor Care Products
- Supplier/Manufacturer of Footwear
- Supplier/Manufacturer of Flooring

Basis of Liability—Owner/Operator of Facility

Duty

The business owner/operator owes the public the duty of exercising ***reasonable care*** in maintaining the premises in a reasonably safe condition, including:

- Proper maintenance of floors,

- Inspecting floors, and
- Removing objects on the floor.

Reasonable Care

In the management of the property, did the owner/operator act as a reasonable person in view of the probability of injury to others given the circumstances of the particular situation. The standard of reasonable care can be established by evidence of:

- Industry Custom or Practice,
- Government Regulations,
  - ✓ OSHA,
  - ✓ ADA,
  - ✓ Building Codes, and
- Voluntary Standards.

Breach of Duty

The business owner/operator can be found liable if the plaintiff can demonstrate that the owner/operator failed to exercise reasonable care and that was the "proximate cause" of the injury (i.e., had it not been for the breach of duty, the individual would not have fallen).

Burden of Proof

Generally, the plaintiff has the burden of proving that the owner/operator did not exercise reasonable care.

If an individual is **injured by slipping on a foreign substance placed or left on the premises by the proprietor or its agent**, the business owner can be liable whether it knows of the dangerous condition or not. In this situation, the plaintiff must prove that the foreign material was related to the defendant's business, and produce some evidence that makes it more likely than not that the defendant was responsible for its existence.

If the offending substance was on the premises through acts of a third person, or if there is no showing of how it got there, the business will generally only be liable if it had actual or constructive knowledge of its presence (i.e., the substance was there for a sufficient length of time so that in the exercise of ordinary care its presence should have been discovered.)

Please note that in establishing constructive knowledge, the burden of proof is generally on the plaintiff. However, a **growing number of jurisdictions such as California and Florida have relieved the plaintiff's burden of proving constructive knowledge.** In general these jurisdictions state that the plaintiff may be relieved of the burden of proving how long a substance was on the floor if he or she can demonstrate that the site had not been inspected within a reasonable time.

Basis of Liability—Manufacturer of Floor Care Products

1. Negligence
2. Strict Liability
3. Breach of Warranty

Basis of Liability—Distributor

1. Negligence (private label distributor—limited jurisdictions)
2. Strict Liability
3. Altered Product

Government Regulations and Voluntary Standards

In regard to maintaining safe walking/working areas, at a minimum, companies must comply with relevant government regulations. Failure to comply with such regulations establishes a company's negligence. However, regulations establish minimum standards of care, while companies and other potential defendants are held accountable for exercising "reasonable care," which is often a higher standard of care than that established by mere regulation.

Consequently, companies should also be cognizant of voluntary industry consensus standards that are relevant and appropriate for the particular circumstances. These standards are often viewed as establishing what the courts consider to be the standard of reasonable care to which a defendant will be held accountable. Moreover, companies and others must view their circumstances and implement other prudent measures designed to reduce the potential for individuals slipping and falling.

This section will address some of the more pertinent regulations and standards and recommend measures that can be taken to minimize potential slip and fall liability.

OSHA

OSHA regulations address workplace safety and are directly relevant to matters involving employees. Additionally, in those situations not involving employees, OSHA regulations may be used as evidence of industry custom or practice. The following is an overview of those OSHA regulations that relate to "walking-working surfaces."

Coefficient of Friction

OSHA recommends that walking surfaces have a static coefficient of friction of 0.5.

General Requirements (29 CFR 1910.22)

- All places of employment, passageways, storerooms, and service rooms shall be kept clean and orderly and in a sanitary condition.

- The floor of every workroom shall be maintained in a clean and, so far as possible, a dry condition. Where wet processes are used, drainage shall be maintained and gratings, mats, or raised platforms shall be provided.
- Every floor, working place and passageway shall be kept free from protruding nails, splinters, holes, or loose boards.
- Aisles and passageways shall be kept clear and in good repair with no obstruction across or in aisles that could create a hazard.
- Permanent aisles and passageways shall be appropriately marked.
- Where mechanical handling equipment is used, aisles shall be sufficiently wide. Improper aisle widths coupled with poor housekeeping and vehicle traffic can cause injury to employees, damage the equipment and material, and can limit egress in emergencies.

Guarding Floor and Wall Openings and Holes (1910.23)

Fixed Industrial Stairs (1910.24)

Portable Ladders (1910.25, 1910.26)

Fixed Ladders (1910.27)

Scaffolding (1910.28)

Manually Propelled Mobile Ladder Stands and Scaffolds (Towers) (1910.29)

Americans with Disabilities Act (ADA)

The Americans with Disabilities Act (ADA) takes into account the findings of the Architectural and Transportation Barriers Compliance Board and when signed into law in 1990, recommended that all accessible routes have a Static Coefficient of Friction (SCOF) of 0.6 and a SCOF of 0.8 for ramps. These recommendations were withdrawn from the ADA in 1999 and no specific ADA requirement has since been adopted.

ANSI Standard for the Provision of Slip Resistance on Walking/Working Surfaces (ANSI A1264.2-2001)

The American Society of Safety Engineers acted as the secretariat for the development of this standard which addresses the concept of "slip resistance" and also sets forth common and accepted practices for providing reasonably safe walking/working surfaces.

Scope

The standard is intended to apply primarily to industrial and workplace situations and is not intended to apply to construction or residential occupancies, or other occupancies such as retail or office facilities where the general public is admitted.

Footwear

In situations where footwear selection can reasonably be controlled, consideration shall be given to slip resistance of footwear to maximize traction, including the required use of special shoes or overshoes when wet conditions are anticipated.

- Foreseeable conditions in the walking/working environment shall be considered in selection of appropriate footwear (i.e., type of flooring, type of tasks to be performed, potential contaminating substances, etc.)
- Footwear traction shall be considered for the foreseeable walking/working tasks to be performed in the workplace. (See Shoe & Allied Trade Research Association Recommendations.) Selection qualities should include:

  - ✓ Slip resistance
  - ✓ Tread design
  - ✓ Tread hardness
  - ✓ Shape of sole and heel
  - ✓ Abrasion resistance
  - ✓ Oil resistance
  - ✓ Chemical resistance
  - ✓ Heat resistance

Mats and Runners

Mats or runners shall be considered for use when walking surfaces do not meet the slip resistance guidelines established by the standard. Mats or runners may be required during wet or inclement weather conditions.

- Mats and runners shall be in areas where it is reasonably foreseeable that operations may encounter slippery contaminants or foreign materials on floor surfaces.

- Mats and runners shall be adequately secured against movement.

- Mats shall be safely installed so that they do not create tripping hazards (they shall have a beveled or flat edge or other appropriate treatment to help reduce the possibility of tripping).

Housekeeping

A housekeeping program shall be implemented to maintain a safe walking/working surface. The standard recommends a written program to ensure consistency. The program should describe materials, equipment, scheduling, methods, and training of those conducting housekeeping.

- The written housekeeping procedures should specify cleaning and maintenance procedures including inspection, immediate response, routine operations, remedial measures and reporting requirements.
- All housekeeping staff, contractors, and others with responsibility for maintaining floors shall be trained regarding the housekeeping procedures including recordkeeping and reporting relating to housekeeping and maintenance.
- Monitoring of areas shall be conducted on a regular basis and shall include: inspection of all working surfaces; notification of persons responsible for clean up; and placement of signage, barriers, etc. until clean up is complete.

Warnings

A warning shall be provided whenever a slip/fall hazard has been identified until appropriate corrections can be affected. Depending upon the circumstances, a warning sign, barricade, or even the placement of personnel may be called for.

Controlled Access

Where an inherently slippery environment exists due to the nature of a product being processed, or otherwise handled, a satisfactory combination of the barricades, containment of spillage and restricted entry shall be utilized.

Selection and/or Treatment of Flooring

Flooring materials should be selected consistent with the ASTM Standard Guide for Selection of Certain Walking Surfaces When Considering Footwear Traction F802-83 (1997) Consideration should be given to replacing the flooring with a material having more pronounced surface asperities. Textured surface coatings shall be considered a viable selection alternative. The material selection should be determined based upon physical testing of the surface with appropriate slip testing devices.

- Where it is not practical to replace flooring, etching, scoring, grooving, brushing, appliqués, coatings, and other such techniques shall be used to provide acceptable slip resistance under foreseeable conditions.

- Occasional testing of surfaces should be performed to monitor slip resistance on walking/working surfaces because some treatments, coatings, etc. deteriorate over time.

Slip Resistance

Walking surfaces should be monitored for their slip resistance characteristics. The standard recommends a slip resistance guideline of 0.5 for walking surfaces in the workplace under dry conditions.

Test methods used to determine compliance with this guideline should have completed approved precision and bias statements in accordance with ASTM E-691-99 (Practice for Conducting Interlaboratory Study to Determine the Precision of a Test Method).

Managing Liability

Now that we have a general understanding of the causes and basis of liability in slip and fall incidents, we can discuss prophylactic measures that are designed to manage or minimize potential liability.

Minimizing Liability of Owner/Operator/Employer

Business owners, operators and employers experience the brunt of liability in slip and fall incidents, and thus have much to gain by taking proactive steps to enhance floor safety and reduce the incidence of slip and fall accidents. The following proactive measures are largely excerpted from a checklist developed by the National Floor Safety Institute located in Bedford, TX, entitled "OSHA Self-Inspection Checklist—Walking and Working Surface Requirements."

Housekeeping

- Establish written procedures
- Keep all work environment, passageways, storerooms, and service rooms clean, sanitary and orderly.
- Keep workroom floors clean and dry. Where wet processes are used, provide drainage, false floors, platforms, mats, etc.
- Keep floors, working areas, and passageways free of protruding nails, splinters, holes, loose boards, or tiles.
- Periodically inspect the walking areas to check for foreign objects, water, and other items that may create falling hazards. Maintain records of your inspection activities including the area inspected, conditions observed, who conducted the inspection, and the time.
- Practice sound recordkeeping by maintaining a cleaning log, including products used, surfaces cleaned, when and by whom tasks are performed, and cleaning procedures.

Employee Training

- Train employees about established safety procedures, cleaning operations, and inspection procedures.
- Provide employees with appropriate product usage training.

- Post written slip and fall prevention and accident handling policies in conspicuous places.
- Keep records of all employee training including individuals trained, subject matter covered, training materials, and date of training.
- Consider a reward system for employees who promote positive safety procedures.

Flooring and Stairs

- Select appropriate flooring materials for the anticipated conditions of use. Select flooring with a SCOF of greater than 0.5 for "high risk" areas.
- Periodic monitoring of slip resistance.
- Inspect flooring surfaces for holes, chips or other trip hazards and make necessary repairs.
- Use non-slip stair treads and landings with abrasive stair nosing.

Matting

- Use absorbent walk-off mats at all doorways that lead to the outside and in other areas where it is foreseeable that slippery conditions exist.
- Use low profile, highly abrasive matting in areas where grease and oil are present.
- Thick mats should be constructed with beveled edges to minimize tripping.
- Use mats with a non-slip backing on wet surfaces.
- Adequately secured against movement.

Cleaning Chemicals

- Maintain "high-risk" areas using a traction enhancing cleaner.
- Select a floor cleaner that enhances slip-resistance and does not leave a slippery soap residue. Rinse thoroughly with clean water after use.
- Select a commercial floor polish with SCOF of 0.5 or higher. Select higher SCOF products for "high risk" areas.

Footwear

- Provide employees with access to slip-resistant footwear and make it a job requirement, when appropriate (ASTM F695).

- Consider foreseeable conditions in the walking/working environment

- Footwear traction

  - ✓ Tread design
  - ✓ Tread hardness
  - ✓ Oil resistance
  - ✓ Chemical resistance
  - ✓ Heat resistance
  - ✓ Shape of sole/heel

Hazard Identification

- Once hazards are identified (i.e., a spill on the floor), post caution signs or barriers preventing access to the spill.

- Post caution signs while mopping.

- Periodically inspect entryways and mop up tracked in rain, snow and debris.

- Eliminate chronic hazards by implementing design changes and conducting frequent equipment inspection and servicing.

- Barricade doors of single entrance rooms when wet mopping.

Minimizing Liability of Manufacturer/Distributor

Quality Control

- Ensure product integrity, consistency and quality so that product is appropriate and safe for its intended use.

- Product testing.

- Maintain records demonstrating product quality.

Education and Training

- Educate and train end user.

- Proper product application.

- General floor safety program.

Warranty

Do not warrant the product to be safe and appropriate for uses inconsistent with its intended use.

REFERENCES

[1] 2002 Liberty Mutual Workplace Safety Index, Liberty Mutual, www.libertymutual.com.

[2] "Fatal Occupational Injuries by Industry and Event or Exposure for 2000," Bureau of Labor Statistics, www.bls.gov.

[3] Leamon, T. B. and Murphy, P. L., "Occupational Slips and Falls: More Than a Trivial Problem," Ergonomics, Vol. 38, No. 3, March 1995, pp. 487-498.

[4] "Slip and Fall Accidents Top List of Employee Injuries," Nation's Restaurant News, August 31, 2000, www.nrn.com.

[5] Report on Injuries in America, 2001, National Safety Council, www.nsc.org/library/rept2000.htm.

[6] Report on Injuries in America, 2001, National Safety Council, www.nsc.org/library/rept2000.htm.

[7] The "Costs of Fall Injuries Among Older Adults," Centers for Disease Control, www.cdc.gov.

[8] Kendzior, R. J., Slip and Fall Prevention Made Easy, Government Institutes, 1999.

Alex Sacher[1]

A Case History: Refutation of Alleged Floor Maintenance Malpractice by the Application of Forensic Biomechanics

---

Reference: Sacher, A, **"A Case History: Refutation of Alleged Floor Maintenance Malpractice by the Application of Forensic Biomechanics,"** *Technology of Floor Maintenance and Current Trends, ASTM STP 1448*, W. J. Schalitz, Ed., ASTM International, West Conshohocken, PA, 2004.

Abstract: This paper is based upon the author's liability expert report prepared for the defendant's attorney who represented a general maintenance contractor. The author was retained to evaluate the technical aspects and merits of the case, and to render an opinion, which would meet the criteria of the Supreme Court's "Daubert"[2] decision.

The complaint charged that the subject (unwittnessed) slip and fall accident was the result of negligent floor maintenance in an office building where the plaintiff was employed. And more specifically, it was due to the defendant's negligent spraying of furniture polish onto a carpeted floor, the accident situs.

A brief review of the circumstances surrounding the accident is presented, followed by: (a) select portions of the plaintiff's "remarkable" (medical terminology for significant and relevant) prior and post accident medical history; (b) the more salient aspects of the plaintiff's deposition testimony and that of the defendant's maintenance supervisor; (c) an inspection of the accident site; (d) coefficient of friction measurements of the subject carpet (before and after application of the plaintiff identified furniture polish) by two independent test laboratories retained by both the defendant and plaintiff; (e) technical considerations (the biomechanics of walking mishaps, the pathophysiologic consequences of injuries and surgeries, et al); and (f) the author's opinion, to wit, that the plaintiff's detailed description of the alleged incident was exquisitely *inconsistent* with the biomechanics of a slip and fall accident. But instead, it is suggested that the accident was actually the result of a stumble, the sequela of the plaintiff's prior chronic locomotor deficiencies.

---

1 President, UPI-Universal Petrochemicals, Inc. 92 Van Ness Court, Maplewood, NJ 07040.

2. C&EN / July 7, 2003, Daubert v. Merrell Dow Pharmaceuticals, Inc. This 1993 case established tough new guidelines for admissibility of scientific evidence in federal courts and told judges they must be gatekeepers and determine what is reliable evidence and what is not. The Daubert Supreme Court decision suggests four questions that judges should consider in determining whether an area or field of science is reliable enough to be used in the courtroom, viz.

i. Is the evidence based on a testable theory or technique?
ii. Has the theory or technique been peer reviewed?
iii. Does the technique have a known error rate and standards controlling its operation?
iv. Is the underlying science generally accepted?

Keywords: slip resistance, coefficient of friction, contaminants, James Machine, horizontal dynamometer pull meter, carpeting, biomechanics.

Introduction

Briefly, the circumstances surrounding the subject "unwitnessed" accident are as follows: the plaintiff alleged that late one morning, she slipped and fell in the "hallway" of an office building where she was employed as an executive secretary for approximately one year prior to the incident. As a result of the incident, the plaintiff claimed to have sustained injuries to her right knee, hip, shoulder, and ankle.

At the time of the incident, the plaintiff was married with two grown children, 47 years of age, 5 feet 3 inches in height, and weighed 150 pounds. She was wearing a pair of flat shoes

Depositions

According to her deposition testimony, the plaintiff arrived at her office between 7:30 am and 8:00 am on the morning of the accident, and carried out her usual duties. At approximately 11:00 am she walked from her workstation "right across" a hallway to her boss' office and that when she "... was approximately two or three inches in front of [and 'in the middle of'] his doorway, [she] slipped." The plaintiff further stated that when she reached this area, "[she] flew through the doorway up in the air, and ... came down on [her] right knee, fell over to the side and hit ... [her] right hip on the floor and put [her] arm, ['right hand'] down to brace [her] fall"; that both of her feet "went behind" her, and both feet came to rest inside the office.

Following the fall, she remained on the floor for approximately ten to fifteen minutes, got up unassisted and returned to her desk. Additionally, she stated that she "... stayed in the office a couple of more hours, putting ice on [her] knee... called [her] orthopedist and left..." driving herself to the doctor's office.

Prior to leaving her office, the plaintiff characterized the alleged "wet spot" (accident situs) on the carpeted floor as "slippery", after testing it with her foot and after running her hand over the surface and "looking at" the residue. Moreover, she stated, "this is furniture polish"--- an opinion she based on "some experience with furniture polish. It was similar to other furniture polish. It was oily, it was slippery." She also stated that the substance on the carpet, which was estimated to be somewhat more that six inches in diameter, was not there the day before; and that she "was the only, the first person to walk into the area that morning."

The Plaintiff went on to explain that she slipped a second time one month later, but did not fall --- outside another executive's office, and that there were yet other slipping incident (without falls) in the same time frame by three other employees.

Although she "investigated" each of these incidents by rubbing her hand on the floor, she never detected a lemon fragrance, "or recalled trying to smell the residue on her hand." No one else, including the other secretary, the building manager, the day porter/custodian, or any of the employees who she alleged had slipped, acknowledged, agreed, or opined that it was furniture polish that made the floor slippery. Finally, other than the alleged "wet spot", she stated that the carpet was not slippery --- a fact borne out not only by her personal experience during approximately one year of employment at the site, but also by the number of other individuals traversing the common area carpeting without a reported slip or slip and fall mishap.

The plaintiff also stated that she took photographs (which were subsequently provided as exhibits) inside the supply room five or six months after her accident. When "I happened to notice that the door ['generally locked'] was wide open one day, I looked in there and saw the various chemicals on the shelf, and on another occasion, I carried my camera with me when the door was open, I took a picture of a furniture polish with lemon oil". She went on to state that, "late one night, [she] saw cans of furniture polish, these same cans, on the cleaning carts" --- an observation she acknowledged was made many months subsequent to the incident.

The maintenance contractor's deposition testimony provided insight into the procedure used to maintain this three-floor building. The night shift operated from 6:00 pm to 10:00 pm with a supervisor overseeing the duties of four maintenance workers. This procedure was in place at the time the accident occurred. He was there at the time of the incident and that "... wood polish ....is only used on the large table in the conference room...[and] for nothing else." The polish used by his firm comes in spray cans readily identifiable by his company's characteristic colors. The maintenance supervisor went on to state that he did not recognize the can of furniture polish identified by the plaintiff in her photograph and that only his own company's furniture polish was "in-use" and no other. Further, they "always put [apply] it [the polish] on top of a rag...and then you put a light passing on top of the table".

Moreover, he stated that there is only one conference room and one conference table on each floor of the building. These tables are not polished every night, but only every once in a while because of the limited use of the conference room...[that] you don't need to do anything [more than] ...take care of it...with a damp cloth...when they're dirty". The workers are not instructed to keep the tables "shiny" – but "just to clean them"; and that normally they employ a feather duster when the surface "doesn't have a lot of dust...and [when dusty] they have a [pretreated] yellow dust cloth...that the dust sticks to"; that doors, in particular, don't get dirty --- just need dusting (dust cloth or feather duster) once in a while.

The supervisor further stated that carpets are vacuum cleaned every night, and once in a while, when dirty, are shampooed (by a company specialist); that the workers "look and make sure there is no garbage [contaminants]" on the floor --- which he inspects every night before leaving; and that he has never received notice or complaints from anyone that the carpet was slippery. The supervisor did comment that the plaintiff's coworkers drink a lot of coffee, and there are coffee spills, and that the stains are cleaned with hot water and a machine. In addition to providing maintenance for the carpeting, the night crew is occasionally asked to mop the grease from the kitchen floors.

Finally, with respect to training, the supervisor stated that he had received instructions from his own supervisor when he first came on the job, and in turn, instructed his workers on "what to do" and "about how to correctly use all the products" --- especially with reference to the application of furniture polish.

Plaintiff's Medical History

The plaintiff's hospital and medical history revealed that 26 years prior to the subject accident, she had suffered a cervical spine injury in an auto accident; that as a result, "she had ...a surgical procedure on her neck." Then four years prior to the subject incident, she sustained an injury to her right knee. According to her physician's notes, "she twisted her knee about a month ago, [and] about a week after that, caught her heel going into the car resulting in a contusion to the knee. Since that time she has had instability of the knee, pain, and clicking."

Because of "exquisite pain", the plaintiff underwent an arthroscopic surgical procedure (partial medial meniscectomy and partial synovectomy) that year and then another arthroscopy the following year --- between the procedures, her knee "had been having stiffness ...[and] would just give way, collapse.... almost every day".

Additionally, according to another physician's report, the plaintiff previously had bilateral podiatric surgery for bunions, and that medically, she suffered from a number of conditions requiring medication, which included treatment for minor deep vein thrombosis.

Physician's notes on the day of the incident indicated that the plaintiff "...slipped on some wet carpet at work. She went down on her left leg, injuring the right knee.... x-rays of the right hip taken in the office today are negative for acute injury. Suggestion of a small bone cyst in the femoral head...x-rays of the right knee demonstrate joint space narrowing on the medial side, early osteoarthritis.... No acute findings on the knee; and that the pain was similar to that of the past."

Because the plaintiff's continued complaints following an arthroscopic procedure four months after the subject incident, she underwent a total knee replacement. According to "Report of Operation", "Preoperative Diagnosis: Osteoarthritis Right Knee...Postoperative Diagnosis: Same".

Inspection of Accident Site

The accident site identified by the maintenance contractor's district manager, was inspected at approximately 11:00 am to coincide with the reported time of the incident. The following observations were made: the hallway and contiguous offices depicted in the plaintiff's photographs were found to be a fair representation of the subject area; the referenced hallway (corridor) where the plaintiff allegedly slipped but did not fall, measured approximately six feet in width; the floor covering in both the corridor and offices consisted of 18" by 18" dark, multicolored, salt and pepper, tufted, looped, nylon carpet with a pile thickness of 0.139 inches (3.5 mm).

The doorway at the end of the corridor (to the left of her boss' office) led to other offices through which a number of individuals passed during the limited time of the inspection. It was noted that this volume of traffic was not concordant with the plaintiff's stated experience that morning, to wit, that she was the first, the only one, to walk through that corridor at 11:00am.

Further, the circulating air temperature and relative humidity were measured and found to be 70 °F and 45 %, respectively. The floors were free of debris, with no visible stains on the carpet, nor dust on windowsills, etc. The district manager reviewed various aspects of the night maintenance crews training, including specific duties and timing for those tasks, which are conducted on a daily basis. Dusting is done with a special, approved, pretreated cloth; furniture is not regularly polished; the application of dust cloth treatment (or furniture polish, if requested) is never applied directly to a surface, but instead, is sprayed onto the cloth in accordance with the manufacturer's instructions. Finally, vacuuming is done daily and is the last duty before leaving the floor.

Dry and Wet Slip Resistance Determinations

An independent test laboratory, recognized by the polish industry, was retained by the defendant (DTL) to determine the static coefficient of friction (SCOF) of both an exemplar and a control (or reference) nylon carpet tile (the same as that on the accident situs floor). The test was conducted in accordance with the ASTM D2047 Standard Test Method for Static Coefficient of Friction of Polish-Coated Floor Surfaces as Measured by the James Machine. The static coefficient of friction of both the control and exemplar carpet tiles was the same. Upon treatment of the carpet with the plaintiff-identified furniture polish, the slip resistance was measured again. The findings showed that whether the carpet was tested "neat" or after a saturation spray, the SCOF values were the same.

In the process of evaluating the slip resistance characteristics of the samples, the carpet was also evaluated for weight changes after a single and also multiple "oversprays" of the furniture polish identified by the plaintiff in her complaint. The weight gain observed following a single "overspray" of furniture polish was negligible (after twelve hours of drying under ambient conditions); and after four "oversprays" the weight gain was only very slight, i.e., 0.028 ounces per square foot or approximately 0.006 ounces for a 6-inch wet-spot as described by the plaintiff. Moreover, the "oversprays" did not produce visual or tactile evidence of a "wet" spot when the subject dark, non-lustrous, multi-colored, nylon carpet was closely examined immediately prior to friction testing (after 12 hours).

The plaintiff retained another test laboratory (PTL) "to determine the coefficient of friction of the submitted carpet...as received...(and also after being) subjected to the plaintiff identified furniture polish". For this purpose, the PTL employed the ASTM C1028-89 Standard Test Method for Determining the Static Coefficient of Friction of Ceramic Tile and Other Like Surfaces by the Horizontal Dynamometer Pull-Meter Method. The testing laboratory used three different test sensors in their evaluation that included, leather, rubber, and "Neolite".

The PTL concluded, "...leather values for treated and untreated carpet do not show a significant difference in slip. The rubber values do show a significant difference in slip...[when the carpet is furniture polish-treated], ...as did the Neolite; and that the slip resistance values...are significantly less than the accepted value of 0.5. Moreover, it was determined ...that the inadvertent application of furniture polish ...would last three (3) weeks, and possible longer..."

Technical Considerations

The following scientific/engineering review, based upon the prevailing state of knowledge and the author's observations and research of more than 40-years, is provided to assist in elucidating the technical aspects of this alleged slip and fall accident; and to disabuse the uninitiate of a number of naïve, misleading and erroneous statements in the PTL report [1-3].

Human biomechanics, is an interdisciplinary science encompassing anatomy, physiology, physics, and mathematics. It is concerned with the interrelation of body structure and function with respect to the kinetics and kinematics of human motion. The forces produced by the body, the forces acting on the body, and the consequences of such forces on tissue deformation, are collectively referred to as the kinetics, whereas the spatial and temporal characteristics are referred to as the kinematics.

The human gait involves complex, integrated neuromuscular skeletal activity which, when disturbed by environmental or purely personal factors, may cause large postural perturbations (irregular movements). These irregular movements, in turn, may lead to an irreversible loss of balance.

A complete walking cycle, (the stride), is from heel contact to heel contact of the same foot (two steps). Within this one second cycle each leg alternates between a stance (supporting) and a swing phase (which represents 40% of the stride). In other words, 40% of the time, one foot is off the ground. Accordingly, walking has frequently been described as a repetitive loss and recovery of balance, and "...a series of catastrophes narrowly averted."

Slip resistance is that property of a surface (footwear or flooring), which denotes its ability to withstand or give protection against an involuntary foot-slide. It is essentially a descriptive term encompassing all the material and human elements that may lead to a slip, and is a function of a number of parameters, among which the coefficient of friction is only one – albeit the most important. Slip resistance is influenced by intrinsic factors, such as an individual's physiological, perceptual, and behavioral condition. Frequently, slip resistance is merely an individual's perception – a subjective, qualitative assessment of the degree to which a particular floor resists the movement of one's shoe sole across its surface, wherein the co-equal contribution of the footwear bottom is either ignored or not even considered.

The standard test method (instrument and protocol) employed to measure coefficient of friction, is dependent upon the product involved and the designated ASTM committee with the direct responsibility (jurisdiction) for its determination --- as follows:

ASTM committee	Jurisdiction	Standard Test Method	Instrument	Sensor Material	SCOF Compliance Criteria
D21	Polishes / Slip Resistance	D2047 {4}	James Machine	Leather	0.5
C21	Ceramic Tile and Related Products	F609 {5}	Horizontal Dynamometer	Neolite	None
F13	Footwear Sole /Heel Materials	F609 {6}	Horizontal Pull Slipmeter	Sole/Heel material	None
F15	Slip Resistance of Bathing Facilities	F462 {7}	Brungraber Mark I	Silastic 382	0.04

It has been shown {8}, {9} that a fall will occur approximately 0.05 seconds after heel contact if the shoe slides forward at the beginning of the stance phase --- but never when it slips backwards at push-off {10}; and further, that the distance of the slide is greater than 4 to 6 inches (10 –15 cm), or the velocity of the sliding foot is greater than 1.6 feet / second, (0.5 m/s). Under these classic conditions, as the sliding foot accelerates forward the same side of the body rapidly falls in a posterior or posterolateral direction. The victim lands squarely on the buttocks or on the side of the buttock, hip, thigh, and shoulder/arm. The final position after impact is either in a sitting or lying position (on the side or supine), depending on distance, velocity, and angle of the foot slide. It should be noted that the body's center of mass always lands close to or directly over the slip situs.

A trip is defined as a sudden loss of footing, the sequela of an interruption of the natural, rhythmic movement of the swinging leg. It occurs at any point where an obstruction impedes or checks the smooth completion of the step, causing a momentary hang-up of the foot. If balance is not restored during the stumbling phase, the body starts to fall in an anterior or anterolateral direction. The reflex and equilibrium reactions are activated, the head and trunk are arched back to counteract the forward momentum, and the arms are abducted to assist in regaining balance. The victim may land prone (arms abducted), sideways or on hands and/or knees with injuries to the head (face)/neck, shoulder/elbow, hands/wrists, hip, thigh, knees, ankle, or foot. Landing distance from the obstacle and the extent of injury are consistent with the body's orientation, vertical force, kinetic energy, momentum, and the hardness of the ground impact site.

A stumble may be defined as an unstable, erratic, arrhythmic and asymmetric foot movement, following a loss of footing and balance due to extrinsic or intrinsic factors. The former may result from a maxi-slip (no fall), trip, collision or push; the latter from fatigue, medication, alcohol, surgery, illness {11}, or age related declines in visual, vestibular, proprioceptive or musculoskeletal functions {12}. A consequence of these intrinsic factors is the atypical or pathological variation from normal walking patterns. All aberrant gait patterns are manifestations of changes in timing, misalignment of body parts, and difficulty in executing movements (slowed reaction times and coordination problems). They can result from responses to pain, muscle weakness or paralysis, spasticity or contactures of muscles, sensory disturbances, and disease {13}. Clearly, any of these conditions may affect the control of the foot during the swing or the stance phase {14} leading to the familiar expression: "tripped over my own foot", "stepped on my own toes", etc., to explain a fall – in the absence of any obstacle to trip over/against. Patently, a stumble will terminate in a fall direction, landing distance, and injury – governed by and consistent with details of its etiology.

Furniture Polish

The principle spray-type furniture polish produced by industry is based on low "solids" (2-5%) silicone or wax emulsions, or combinations of both, with or without lemon oil. According to the manufacturer's technical literature for the plaintiff identified furniture polish, it is a thin white emulsion with a lemon fragrance. The product contains "a rich blend of silicones, cleaners, and lemon oil", and a caveat that states, "Avoid overspray as surface may become slippery".

Lemon oil, according to the Merck index {15}, consists primarily of dipentene terpenes, and may be used as a wetting and dispersing agent; Boiling Range of 150° – 175° C. The speed of evaporation is 25 minutes (according to Union Carbide {16}, ...expressed in terms of minutes required for 0.5 cc ....to evaporate from filter paper – (which) will vary widely according to atmospheric conditions, temperature, humidity, barometric pressure, etc.").

Anatomy / Injuries / Locomotor Consequences

The following subject matter is derived in part by quoting, or paraphrasing information found in various tests {17} and the Merck Manual {18}.

The knee joint is the largest and most complex joint in the human body. Placed midway in each supporting column of the body, it is subject to severe stresses and strains in its combined functions of weight bearing and locomotion, and is probably the most susceptible to injury of any of the joints.

The menisci are circular rims of fibrocartilage situated on the articular (knee joint) surfaces of the head of the tibia, and serve in a shock-absorbing capacity. The medial cartilage is shaped like a large letter C, and is not freely movable because of its secure anchorage to the tibial collateral (medial) ligament at the medial side of the knee. Largely because of these points of attachment, the medial cartilage is more frequently injured than the lateral.

When the knee receives a blow from the side or when it is subjected to severe wrenching in the weight-bearing position, it can easily be forced beyond its normal range of rotary motion, resulting most likely in a tear of the ligament on the opposite side. If the force is great enough to be transmitted to the deep layer of the tibial collateral ligament, it is likely to affect the medial meniscus, which is attached to the ligament.

Removal of living tissue may alter the functioning of the human body, and is most noticeable with respect to knee surgery. Removal of half a meniscus at the knee joint ultimately results in degeneration (osteoarthritis) of the joint, and the degree of degrade is found to be directly proportional to the amount of menisci removed.

Osteoarthritis (OA) is a degenerative joint disease wherein all tissues in and around involved joints are hypertrophic (excessively enlarged). Primary OA includes peripheral joints, e.g., the 1st metatarsal phalangeal joint (big toe), hip, knee, etc. Although etiology is unknown, many mechanisms can initiate this event, such as acute or chronic trauma (including fracture) to the cartilage or tissue surrounding it. Two pathologic processes characterize OA: deterioration and loss of a bearing surface, and proliferation of all osteoarticular tissue at the margins of the joint and under the detached joint surface. Initially, OA is noninflammatory and onset is subtle and gradual, usually involving one or only a few joints. Pain is the earliest symptom. As the disease progresses, joint motion becomes diminished, flexion contractures occur, and tenderness and crepitus or grating sensations appear. Joint enlargement is induced by the proliferative reactions of cartilage, bone, ligament, tendon, capsules, and chronic synovial proliferation. Then, as the ligaments become lax, the joint has increasing instability with more local pain.

The pathophysiologic process usually proceeds to the appearance of symptoms and signs, varying degrees of disability, and functional compromise. Total joint replacement should be considered only when all conservative therapy has failed.

Bursitis is an acute or chronic inflammation of a bursa. Bursa are saclike cavities filled with synovial fluid (a transparent viscid lubricating fluid) and located at tissue sites where friction occurs. Most bursitis occurs in the shoulder, but other common forms exist, e.g. at the first metatarsal head (bunion). The etiology of most bursitis is unknown, though it may be caused by trauma, chronic overuse, inflammatory arthritis, etc.

Metatarsalgia, a general term for pain over the ball of the foot, usually is the result of injury to the metatarsophalangeal articulation. Pain involving the metatarsophalangeal joint is a common occurrence, almost entirely due to misalignment of the joint surfaces, causing subluxations (partial dislocation of one of the bones in the joint) with eventual destruction of joint cartilage (degenerative joint disease). Such subluxations are seen in patients with hallux valgus deformity (bunion). Osteoarthritis of the first metatarsophalangeal joint is extremely common. Initially the only sign may be slight swelling of the joint due to capsular thickening. The joint is tender, and shoes aggravate the condition. As the condition worsens, pain increases and osteocartilaginous exostoses (benign bone tumor) formation begins to limit joint motion; the patient no longer bends the joint during walking. In cases recalcitrant to conservative therapy, surgery may be necessary.

Conclusions

The author's professional opinions based on a reasonable degree of scientific and engineering certainty are as follows:

1- The PTL's naïve nexus of the 0.5 static coefficient of friction (SCOF) compliance criterion "as a measure of a safe walking surface", in relation to the method (ASTM C1028) employed by it to determine the SCOF of the subject carpet, was meaningless, inappropriate, and misleading.

2- ASTM C1028, the standard test method for "Determining the Static Coefficient of Friction of Ceramic Tile and Other Like Surfaces by the Horizontal Dynamometer (under the jurisdiction of ASTM Committee C21 on Ceramic Whitewares and Related Products), has no compliance criterion, not 0.5 or any other numerical value; and patently not for the carpeted surface.

3- The 0.5 SCOF compliance criterion is, in fact, derived from a correlation of "field" experience with James Machine SCOF determinations, and is not a universal reference value; that instead it is only meaningful when measurements are made in accordance with ASTM D2047 (under the jurisdiction of Committee D21 on Polishes and Subcommittee D21.06 on Slip Resistance). The method is also specified by both Industry (Consumer Specialty Products Association){19}, and Government (General Services Administration) {20}.

4- There is no consistency between the SCOF James Machine values obtained when comparing the test results of ASTM D2047 with ASTM C1028, or, in fact, any of the four methods previously listed {21}{22}.

5- The PTL statement "...that a COF value of 0.5 was *established* by the Federal Trade Commission in 1953 {23}" is erroneous. The FTC merely published "Proposed Trade Practice Rules for the Floor Wax and Floor Polish Industry" (at the behest of the Floor Polish Industry) in re the "Improper Use of (various) Terms, e.g. 'Slip Proof', 'Non Slip', 'Slip Resistant', Slip Retardant', etc."); and "...that for the purpose of compliance with this Rule, the product shall have a coefficient of friction of not less than 0.5 as determined by the test for slip resistance as used by the Underwriters' Laboratories, Inc. [The James Machine]..."

6- Whereas the coefficient of friction may be defined as the intrinsic property of two interfacing, interacting solid surfaces at the slip-plane – a two-dimensional phenomenon --- the friction developed with carpeting is three-dimensional. In this instance, the foot, under a vertical component of force (120% of the individual's body-weight), literally penetrates the irregular (sculptured), textured (looped or cut) and pliant (deep or short pile) surface of the carpet. To date, none of the available friction testers adequately measure the total friction (resistive force) developed between the footwear and carpeting --- the friction produced by the three-dimensional, mechanical interlocking of the rear outer-edge of the heel at "heel strike" (contact), at "mid-stance" while the foot rocks forward from heel-to-toe, and at "push-off" by the toe/forefoot, beneath the carpet surface --- which would account for the negligible amount of reported slips and falls on a carpet, per se.

7- As a corollary, the above is probably the reason that no friction tester, test method or compliance criterion has been recognized, approved or specified by the Carpet and Rug Institute.

8- The defendant testing laboratory used the James Machine not to determine compliance with the SCOF of 0.5, but solely for the purpose of monitoring the effect of an "overspray" on the SCOF of the untreated carpet. The DTL was thereby able to show that the SCOF of the carpet, whether tested "neat" or following a single or multiple "overspray" remained unchanged.

9- These results are not surprising since the primary function of carpet mats at building entrances "...is to stop, catch, trap, hold, and hide dirt, grime, sand, and water" --- according to ISSA {24}.

10- One square yard of commercial grade carpeting can hold (accumulate) up to one gallon of water and up to five pounds of dirt, grime, etc. --- according to 3M {25}.

11- The carpet "weight gain" following a single or multiple "overspray" of the plaintiff-identified furniture polish, and the results of the DTL's "weight retention" measurements after twelve hours of "drying" under ambient conditions, do not remotely support the plaintiff's allegation or the PTL's "determination" with respect to the "lasting" quality of a single "overspray" of the furniture polish on the subject carpeting.

12- That despite the PTL's determination "...that the inadvertent application of the furniture polish ... would last three weeks....", the plaintiff testified that she never detected a lemon fragrance (nor did any of her co-workers) on the morning of her falling accident --- approximately twelve hours after the night maintenance crew departed, even though she rubbed her hand over the wet carpet surface and examined the *residue*.

13- The "overspray" caveat on the plaintiff-identified furniture polish container was clearly with reference to hard, non-porous, non-wicking flooring (e.g. ceramic tile, varnish coated wood, resilient tile, marble, terrazzo) --- not carpeting.

14- A simple, single "overspray" of the lemon oil furniture polish emulsion, would readily and rapidly be adsorbed on the surface of the fibers and dispersed within the carpet, and then evaporate well within the twelve hour hiatus between the alleged "overspray" by the night maintenance crew before leaving at 10:00 pm the night before, and the plaintiff's accident at 11:00 am the following day.

15- With respect to terminology, the PTL overlooked the difference ---between *spraying* a surface, which is a deliberate act with the intention of depositing a given amount of material on a surface, and an accidental overspray wherein some very small portion of the spray may have spread beyond the end(s) of the polishing rag or even the edge of the table. In this instance, the amount of material deposited would be negligible after 12 hours.

16- The excellent traction provided by the interaction of the high friction wool-like carpeting, with the flat (low heels), composition rubber-bottom footwear worn by the plaintiff, would significantly reduce the probability of a slip and fall.

17- The custodial services provided by the outside contractor including the training, supervision, and monitoring of the cleaners' workmanship were in accord with the traditional standard practices of the industry.

18- The plaintiff's detailed description of her fall, is exquisitely *inconsistent* in every respect with the biomechanics of a slip and fall accident, i.e. the direction of her slipping foot: (both) went out behind her; the direction of her fall: went flying forward "up" in the air through the doorway; the ground and body impact sites: landed in the office a number of feet from the door (legs and feet included), and came down onto her right knee, with her arms outstretched to brace her fall; and the nature of her injury: swelling and abrasion of the right knee, tenderness around her right hip, and right shoulder pain --- as opposed to her foot/feet going out in front of her, with her body falling backwards (in a posterior or posterolateral direction), landing directly over/adjacent to the slip situs (the "wet" spot which was in the hallway, "two to three inches anterior to her boss' door"), and impacting her buttocks, side of the hip, thigh, etc. --- but not onto her knee. That, in fact, the description of her accident is consistent in every detail with a *trip* or a stumble --- with the former highly unlikely in the absence of an obstacle or an obstruction to trip over.

19- Further, it is the author's professional opinion that the proximate cause of the plaintiff's falling accident and attendant pain (was not according to her physician's notes), a derivative of an "acute injury", a fracture, or a torn tendon, ligament, or cartilage, but was instead, from a "contusion ...swelling, and ...abrasion of her right knee"; and moveover, that it was the sequela of a stumble and loss of footing --- stemming from purely personal factors, such as her prior injuries and surgeries (resulting in slowed reaction times, coordination and balance problems), fatigue, medication, and /or inattention all contributing factors, according to the National Safety Council in falling accidents, which cannot be readily identified or controlled.

20- And finally, that the plaintiff's serious, personal (pre-existing, physical, and medical) problems, i.e. space-narrowing of her right knee ("early osteoarthritis"), two arthroscopic surgeries (meniscectomies), bilateral podiatric surgeries, deep vein thrombosis, etc., markedly affected her ability to ambulate in a care-free manner, or to recover balance and equilibrium following a misstep; and that her "carpet – made slippery – by furniture polish" rationale, is without merit, and clearly a reassignment of *blame*.

References

[1] Sacher, A., "Slip Resistance and the James Machine 0.5 Static Coefficient of Friction – Sine Qua Non," *Standardization News,* August 1993, pp. 52-59.

[2] Sacher, A., "Retrospective Identification of Misstep Type by Analysis of Falling, Landing and Injury Details", Proc. 13th South. Biomed. Eng. Conf., Biomed Eng. Recent Develop, 1994, pp.571-4 Univ. of District of Columbia, Washington, DC.

[3] Sacher, A., "The Application of Forensic Biomechanics to the Resolution of Unwitnessed Falling Accidents," *Journal of Forensic Science*, September 1996, pp. 776-81.

[4] ASTM D2047, Standard Test Method for Static Coefficient of Friction as Measured by the James Machine, *Annual Book of ASTM Standards, Vol. 15.04, American Society for Testing and Materials, 100 Barr Harbor Drive, West Conshohocken, PA 19428-2959.*

[5] ASTM C1028, Standard Test Method for Determining the Static Coefficient of Friction of Ceramic Tile and Other Like Surfaces by the Horizontal Dynamometer Pull Meter Method, *Annual Book of ASTM Standards Vol. 15.02, American Society for Testing and Materials, 100 Barr Harbor Drive, West Conshohocken, PA 19428-2959.*

[6] ASTM F609 Standard Test Method for Static Slip Resistance of Footwear, Sole, Heel, or Related Materials by Horizontal Pull Slipmeter (HPS), *Annual Book of ASTM Standards, Vol. 15.07, American Society for Testing and Materials, 100 Barr Harbor Drive, West Conshohocken, PA 19428-2959.*

[7] ASTM F462 Standard Consumer Safety Specification for Slip – Resistant Bathing Facilities, *Annual Book of ASTM Standards, Vol. 15.07, American Society for Testing and Materials, 100 Barr Harbor Drive, West Conshohocken, PA 19428-2959.*

[8] Perkins, P.J., "Measurement of Slip Between the Shoe and Ground During Walking". Walkway Surfaces: Measurement of Slip Resistance, ASTM STP 649 (Carl Anderson and John Senne, editors), ASTM, Philadelphia, PA 1978, pp. 71-87.

[9] Strandberg. L. and Lanshammer. J., "The Dynamics of Slipping Accidents", Journal of Occupational Accidents, 1981, pp. 153-162.

[10] Strandberg. L. and Lanshammer. J., "On the Biomechanics of Slipping Accidents," Biomechanics VIII-A (H. Matsui and K. Kobayashi, editors), Human Kinetics Publishers, Champaign, IL, 1983, pp. 397-402.

[11] National Safety Council, "Falls on Floors", Data Sheet I-492, Itasca, IL.

[12] Tideiksaar, R., "The Biomechanics and Environmental Characteristics of Slips, Stumbles, and Falls in the Elderly", B.E. Gray, Editor, Slips, Stumbles, and Falls Pedestrian Footwear and Surfaces, West Conshohocken, PA, American Society for Testing and Materials, ASTM STP 1103; 1990, pp. 17-27.

[13] Adrian, M.J., "Biomechanics of Human Movement", Benchmark Press, Indiana, 1989.

[14] Winter, D.A., "The Biomechanics and Motor Control of Human Gait: Normal, Elderly and Pathological", 2nd Ed, Ontario, Univ. of Waterloo Press, 1991.

[15] The Merck Index, 8th Edition, Merck & Co., New Jersey 1968.
[16] Solvent Selector, Union Carbide Chemicals Company, New York.
[17] Luttgens, K., Deutsch, H., Hamilton, N., "Kinesiology, Scientific Basis of Human Locomotion", 8th Edition, Brown & Benchmark, Iowa, 1992.
[18] The Merck Manual of Diagnosis and Therapy, Berkow, R. Editor in Chief, 14th Edition, 1982.
[19] "CSMA Procedures for Products Classified as to Slip Resistance", CSMA Bulletin 308-70 Dec. 8, 1970, Chemical Specialties Manufacturers Association, formerly, now CSPA, Washington, D.C.
[20] "General Services Administration, Federal Specification, Finish, Floor, Water-Emulsion", P-F-430C, Oct. 26, 1978.
[21] Sacher, A., "Is the 0.5 Static Coefficient of Friction Value a Benchmark or a Watershed?", Ceramic Engineering Science Proceedings, Vol. 13, Nos. 1-2, 1992, pp. 29-45.
[22] Sacher, A., "Use, Misuse, and Abuse of the James Machine 0.5 Static Coefficient of Friction - Epilogue: American with Disabilities Act Accessibility Guidelines. In: Sacher, A., chairman, Owens, J.M., co-chairman, International Symposium on Slip Resistance: The Interaction of Man, Footwear, and Walking Surfaces, 1995, Oct. 30-31, NIST/Gaithersburg, MD.
[23] Federal Register, Tuesday March 17, 1953, p.1513.
[24] "Mats and Matting", ISSA Educational and Training Aids", International Sanitary Supply Association, Illinois.
[25] Berwin, H., 3M, Private Communication, 1995.

Mark A. Gindling[1]

Regulatory Issues Affecting Floor Polish

Reference: Gindling, M. A., "Regulatory Issues Affecting Floor Polish," Technology of Floor Maintenance and Current Trends, ASTM STP 1448, W. J. Schalitz, Ed., ASTM International, West Conshohocken, PA, 2004.

Abstract: This paper includes three sections. The first, "Crosslinked Floor Finishes (Zinc Metal)," reviews a position paper written by the Polishes and Floor Maintenance Division of CSPA (CSMA). Allowable zinc discharge amounts are regulated by the EPA through local POTW's (sewer districts), and are based on the receiving stream's natural zinc content. The second, "Maintenance Procedures For Flooring Containing Asbestos," reviews a study done by the Polishes and Floor Maintenance Division of CSPA (CSMA) and the U.S. EPA. Review recommended procedures, resulting from the study, for stripping, buffing and high speed-burnishing asbestos containing floors. The third, "Americans With Disabilities Act, Accessibility Guidelines, Section. A4.5 Ground And Floor Surfaces Appendix B Slip Resistance Requirements," reviews work done by the Polishes and Floor Maintenance Division of CSPA (CSMA) and the U.S. Architectural and Transportation Barriers Compliance Board.

Keywords: acrylic sealer, asbestos, Bulletin #4, floor maintenance, James Machine, metal crosslinked, static coefficient of friction, POTW, Zinc

Crosslinked Floor Finishes (Zinc Metal)

In late 1980, the Polishes and Floor Maintenance Division of the Chemical Specialties Manufacturers Association (CSMA) now (CSPA) the Consumer Specialty Products Association became concerned that the U.S. Environmental Protection Agency may ban the use of zinc in various end-use products including floor polishes. For many years the floor polish industry had relied on the use of zinc to chemically crosslink polymers used to make floor polishes. The use of zinc helps provide many useful benefits to the modern day floor polish by providing detergent resistance for daily cleaning yet allowed for easy removal when it became time for the floor polish to be removed from the floor.
The regulatory problem with zinc comes from the fact that zinc is found on the EPA 65 Toxic Pollutants and the 126 Priority Pollutants list of the Clean Water Act. This forces some POTW (publicly owned treatment works-sewer districts) to enforce a limit on the amount of zinc that can be discharged down a sanitary sewer. The limit is set based on the amount of zinc that naturally occurs in the POTW's receiving stream (river or lake). Some areas of the country have no limits while others have very low limits (as low as 1 ppm). The regulated community of end-users varies from sewer district to sewer district.

The regulatory problem with zinc comes from the fact that zinc is found on the EPA 65 Toxic Pollutants and the 126 Priority Pollutants list of the Clean Water Act. This forces some POTW (publicly owned treatment works-sewer districts) to enforce a limit on the amount of zinc that can be discharged down a sanitary sewer. The limit is set based on the amount of zinc that naturally occurs in the POTW's receiving stream (river or lake). Some areas of the country have no limits while others have very low limits (as low as 1 ppm). The regulated community of end-users varies from sewer district to sewer district.

The industry embarked on an aggressive research project to replace zinc, while maintaining the properties that zinc provided. The result of this research provided a number of zinc free polymers for formulators to choose from, but none gave the ultimate durability that the zinc polymers could provide.

For those who need a zinc free floor polish, most major manufacturers have a product that will provide an adequate level of performance. However, most floor polishes on the market today contain zinc.

The Polishes and Floor Maintenance Division of the Chemical Specialties Manufacturers Association (CSMA) now (CSPA) developed a "Position Paper On The Use Of Zinc in Floor Finishes" to help explain to regulators and news media why zinc is important to our industry and why we have developed zinc-free floor finishes.

Maintenance Procedures For Flooring Containing Asbestos
Recommendation for Asbestos Flooring Based on EPA/600/SR-95/121

In August of 1995, a five-year study involving the maintenance of asbestos flooring by the U.S. EPA, the Chemical Specialties Manufacturers Association and the International Sanitary Supply Association came to a close. The EPA published the findings of the study in the Research & Development document EPA/600/SR-95/121. The following recommendations are based on this study.

Preparation Of Floors:

1. Apply stripping solution so that the floor remains wet during the entire operation.
2. Use the least aggressive stripping pad (green) or brush to do the job.
3. Use the lowest speed floor machine possible to do the job (175 rpm).
4. Pick up stripping solution and dispose down sanitary sewer.
5. Rinse floor thoroughly.
6. Apply two coats of an acrylic sealer that resists the action of normal floor finish strippers.

7a. Floors that are to be spray buffed or have no buffing done to them, apply a minimum of two coats of floor finish.

7b. Floors that are to be high speed buffed (1000 rpms or greater), apply a minimum of four coats of floor finish.

8. Scrub and recoat floors when they show signs of wear. A good rule of thumb for recoating in buffing operations is, a coat of finish is removed after three to four buffings. Recoat buffed floors based on your buffing time schedule.

9. When it becomes time to strip the floors again, following steps 1 through 5 using a stripper that is not designed to remove the acrylic sealer. In step 6 apply only one coat of seal if the seal looks worn. Then proceed with step 7a or 7b.

A copy of the Research and Development document EPA/600/SR-95/121 can be obtained from the National Technical Information Service, 5285 Port Royal Road, Springfield, VA 22161. Telephone number 703-487-4650.

Americans With Disabilities Act, Accessibility Guidelines, Section. A4.5 Ground and Floor Surfaces Appendix B Slip Resistance Requirements

In the spring of 1993, members of a task force, from the Chemical Specialties Manufacturers Association (CSMA) now Consumer Specialty Products Association, (CSPA) chaired by Mark Gindling, Director of Research for Buckeye International, Inc., met with members of the U.S. Architectural and Transportation Barriers Compliance Board to discuss the 0.6 static coefficient of friction for level walkway surfaces, and 0.8 static coefficient of friction for ramps, found in Appendix B of 28 CFR Part 36 Standards for Accessible Design Americans with Disabilities Act Accessibility Guidelines (ADA).

From that meeting, the U.S. Architectural and Transportation Barriers Compliance Board agreed to revise its Bulletin #4 to explain how the numbers were chosen. The bulletin goes on to further explain how the use of just any portable slip-testing machine is not acceptable for measuring floors to see if they meet these suggested static coefficient of friction limits. It further explains the use of the 0.5 static coefficient of friction as measured by ASTM D-2047 Standard Test Method for Static Coefficienet of Friction of Polish-Coated Floor Surfaces as Measured by the James Machine, which is used by the floor finish industry to ensure that our products are acceptable for use on walkway surfaces.

In April 1995, the CSMA Task Force met with a committee that is in charge of reviewing the entire 28 CFR Part 36 Standards for Accessible Design Americans with Disabilities Act Accessibility Guidelines. At that meeting, task force members gave testimony, that explained why the 0.6 and 0.8 static coefficient of friction suggested limits should be removed from Appendix B and from the law itself. The Task Force was asked to write a letter giving all the technical reasons why this should happen so that it could become part of the permanent record.

Representatives from the Justice Department were present for the testimony and agreed to refrain from enforcing this part of the standard until the standard could be rewritten.

These changes were made in the proposed rule making found in 36 CFR Parts 1190 and 1191 in the Federal Register on November 16, 1999.

Final testimony was made to the U.S. Architectural and Transportation Barriers Compliance Board on March 13, 2000 to reaffirm the changes proposed in the November 16, 1999 Federal Register.

As of July 23, 2002 the Board expects the final rule to pass some time at the end of the year 2002.

References

"Position Paper On The Use Of Zinc," CSPA, 900 17th St. NW, Suite 300, Washington, DC 20006.

"Research & Development Document EPA/600/SR-95/121," National Technical Information Service, 5285 Port Royal Road, Springfield, VA 22161.

Personal correspondence, Mr. Brian Black, Assistant Director Buildings Codes & Standards, EPVA, 111 West Huron Street, Buffalo, NY 14202, CSPA, 900 17th street NW, Suite 300, Washington, DC 20006.